RECENT DEVELOPMENTS IN TROPICAL CYCLONE DYNAMICS, PREDICTION AND DETECTION

Edited by **Anthony R. Lupo**

Recent Developments in Tropical Cyclone Dynamics, Prediction, and Detection

http://dx.doi.org/10.5772/61455
Edited by Anthony R. Lupo

Contributors

Jordan Rabinowitz, Jagabandhu Panda, Kasturi Singh, Krishna K. Osuri, Naresh Vissa, Yuriy Kuleshov, Andrew Mercer, Alexandria Grimes, Jiayi Pan, Yujuan Sun, William Perrie, Song Yang, Joshua Cossuth, Qinglan Li, Hongping Lan, Johnny C.L. Chan, Chunyan Cao, Cheng Li, Xingbao Wang, Wen-Zhou Zhang, Sheng Lin, Xue-Min Jiang, Jinwoong Yoo, Robert Rohli, Jennifer Collins, Yi-Ting Yang, Hung-Chi Kuo, Eric Hendricks, Melinda Peng, Kevin K. W. Cheung, Guoping Zhang

CBS, Edition **2018**

Published by InTech

Janeza Trdine 9, 51000 Rijeka, Croatia

Publishing Process Manager Romina Rovan

Technical Editor SPi Global

Cover Designer InTech Design Team

Additional hard copies can be obtained from orders@intechopen.com

Recent Developments in Tropical Cyclone Dynamics, Prediction, and Detection, Edited by Anthony R. Lupo
p. cm.
Print ISBN 978-953-51-2702-4
Online ISBN 978-953-51-2703-1

Contents

Preface

There has been a large amount of research on tropical cyclones, one of nature's most destructive phenomena. This research has resulted in recent improvements for forecasting the arrival, path, and intensity of these storms. Today, however, tropical cyclones can still strike with little warning. Tropical cyclones continue to bring destruction, as well as disruption, to societies that are exposed to their threat. These storms cause as many problems for developed nations as for less developed nations, a good example being Hurricane Katrina (2005).

Of course RADAR and satellite technology and numerical models continue to increase our understanding of the structure and dynamics of tropical cyclones. The monitoring of tropical systems continues to improve in all the world's ocean basins. We are now at the point of understanding the climatological occurrence of tropical cyclones and also the interannual and interdecadal variations in their occurrences. We can also see climatological trends in tropical cyclone occurrence for each ocean basin and can make reasonable assessments about the frequency of its occurrence in the future. Concern about climate change and the change in frequency of extreme events often invoke projections about how tropical storm intensity may change as well. As stated in the book"Recent Hurricane Research - Climate, Dynamics, and Societal Impacts", we understand that these powerful storms represent the cooperative interaction between atmospheric and oceanic conditions and that they will occur only when conditions in both are favorable.

The book represents a compilation of more cutting-edge research on tropical cyclones and their impacts from researchers at many institutions around the world. Each contribution has been reviewed. This book has been divided into three sections, and each is organized by topic. The first section, Tropical Cyclone Dynamics, contains contributions that explore the physical factors contributing to storm development and maintenance. This includes the use of traditional diagnostics as well as statistics in explaining their occurrence. The first chapter examines the impact of tropical cyclone occurrence on the characteristics of the Western North Pacific. Traditionally, studies have examined the impact of the ocean on tropical cyclone characteristics. The second chapter explores the influence of the synoptic scale on the rapid development of tropical cyclones using principal component analysis. The outcome gives insight into which variables are important in forecasting rapid development.

The next chapter in section 1 performs a thorough analysis of the synoptic considerations for understanding tornadogenesis in land-falling storms and the improvement of recognition of these conditions by the operational community. The dynamics and biological response of the upper ocean are the subject of the fourth chapter in this section. Examining a case study (Typhoon Cimaron, 2006), the authors demonstrate that the ocean is restored to the pre-storm equilibrium in two to three weeks. They also demonstrate that phytoplankton blooms can occur within days of a storm's passage. The next chapter examines the planetary-scale low-level dynamics in the rapid development of Hurricane Wilma (2005). The authors show that large-scale low-level circulations associated with midlatitude troughs play an important role in rapid tropical cyclogenesis. The final chapter in this section explores the interaction of mesoscale convective systems and latent heat release associated with deep convection in the rapid development of two tropical cyclones.

The second section presents the use of satellite-based remote sensing in the detection and climatology of tropical cyclones. The first chapter presents a review of satellite-based techniques in examining the location, structure, and intensity of tropical cyclones using many different platforms. This also includes a discussion of their detection and their precipitation patterns. The advent of remote sensing has greatly helped to reduce the ability to forecast these events. The second chapter presents a satellite-based climatology of concentric eyewall occurrence in tropical cyclones. They looked at the frequency with which eyewall replacement happened with the inner wall relative to the outer wall, as well as the maintenance of concentric eyewalls for some time. They find that the inner (outer) eyewall replacement is associated with internal (environmental) dynamics. They find significant variability related to El Niño and Southern Oscillation in the West Pacific Ocean Basin, but no corresponding variability in the Atlantic.

The topic of the third section is contributions to modeling tropical cyclones as well as their prediction. The first chapter presents an in-depth look at the climatology of Northern Indian Ocean region cyclones. Detailed records for this region are less than 40 years old. The authors examine the occurrence, movement, structure, and intensification of tropical cyclones in the region. They also look at the relative role of the atmospheric and oceanic influences on these, with the goal of improving predictability in this region of the world. The second chapter examines the occurrence and prediction of heavy rainfall associated with land-falling tropical cyclones in the southwest part of the West Pacific Ocean Basin. The authors use nonparametric statistical methods. They develop and test statistical models of the probability of heavy rainfalls and show the success of their methods in improving forecasts. The last chapter uses statistical techniques to model the behavior of tropical cyclone activity in the Southern Hemisphere. Using linear multivariable and support sector regression techniques, the authors demonstrate they can reasonably model the regional annual and temporal occurrence of tropical cyclones. These methods could potentially improve current forecast techniques.

In closing, the research presented here adds to the current database on what is known about tropical cyclone behavior and has the potential to lead to the development of better forecasting methods, which will save lives and property. This book would make a great supplement to any course on tropical meteorology, which highlights current research. I would like to thank all those that made this book possible including the publication staff at InTech publishers and the authors who chose to publish their work in this volume.

Professor Anthony R. Lupo
University of Missouri
Columbia, MO, USA

Section 1

Tropical Cyclone Dynamics

Chapter 1

Influence of Tropical Cyclones in the Western North Pacific

Wen-Zhou Zhang, Sheng Lin and Xue-Min Jiang

Additional information is available at the end of the chapter

http://dx.doi.org/10.5772/64009

Abstract

The Western North Pacific (WNP) is the most favorable area in the world for the generation of tropical cyclones (TCs). As the most intense weather system, TCs play an important role in the change of ocean environment in the WNP. Based on many investigations published in the literature, we obtained a collective and systematic understanding of the influence of TCs on ocean components in the WNP, including sea temperature, ocean currents, mesoscale eddies, storm surges, phytoplankton (indicated by chlorophyll *a*). Some ocean responses to TCs are unique in the WNP because of the existence of the Kuroshio and special geographical configurations such as the South China Sea.

Keywords: tropical cyclone, typhoon, influence, ocean response, Western North Pacific

1. Introduction

The Western North Pacific (WNP), including its marginal seas (similarly hereafter), is the most favorable area in the world for the generation of tropical cyclones (TCs) since more TCs are born in this area than in any other region every year. The TCs in the WNP account for about one-third of TCs born in the world oceans [1, 2]. According to the data during the period of 1971–2000, the annual generation rate of TCs is 27.2 in the WNP [2]. For the activity of TCs, the South China Sea (SCS) is the most important marginal sea of the WNP where many TCs pass through and some locally form every year. Averaged from 1968 to 1988, the annual mean number of TCs in the WNP is 25.7 among which the numbers of the TCs passing over the SCS and the TCs forming in the SCS are 10.3 and 3.5, respectively [3]. The TCs in the WNP can occur in any month of the year but most of them are generated in the boreal summer (June-August) and autumn (September-November). Strong TCs are traditionally called typhoons in the WNP.

As the most intense weather system which usually causes drastic air-sea interaction, TCs play an important role in the change of ocean environment in the WNP and attract many researches focusing on this topic. By reviewing these researches, we aim to get a collective and systematic understanding of the influence of TCs in the WNP.

2. Temperature response to TCs

2.1. In the open ocean

Sea surface temperature (SST) response to TCs has been widely noticed and extensively investigated in the world oceans (e.g., [4, 5]). TCs usually cause SST to drop (SST cooling) via vertical mixing, air-sea heat flux, and advection. Vertical mixing is regarded as the most important mechanism of the cooling within the initial mixed layer and meanwhile it is responsible for the warming below the initial mixed layer [4]. Owing to the asymmetry of both wind stress and its rotation with time, the strongest SST cooling is shifted to the right of TC track in the Northern Hemisphere [4, 6]. In addition, this rightward shift or bias depends on the TC translation speed [4, 7]. Upwelling (vertical advection) enhances the SST cooling associated with TCs, especially slowly moving ones, and tends to reduce its rightward shift [4]. The exclusion of upwelling may result in the underestimation of the SST cooling as shown in Chiang et al. [8]. The cooling caused by upwelling may overwhelm the warming due to vertical mixing, resulting in a subsurface cooling, in the subsurface layer where the vertical gradient of temperature is large [6].

Mei et al. [9] demonstrated that stronger, slower-moving, and higher-latitude TCs usually cause a larger magnitude of SST anomaly. Slowly moving typhoons tend to induce larger SST drop than fast moving typhoons for the former have relatively longer residence time, producing a stronger vertical mixing and more heat loss from ocean to air. The magnitude of the SST cooling depends not only on the intensity and translation speed of typhoon, but also on the preceding thermal structure (i.e., mixed layer depth and upper-ocean stratification) of the upper ocean. Lin et al. [5] showed that a typhoon causes quite smaller SST cooling under the condition with a thicker warm upper layer than it does under the climatological condition. The scope of the SST cooling is related to the typhoon size. D'Asaro et al. [7] observed that among several TC-induced cold wakes, the smallest typhoon, Megi in Philippine Sea, produced a narrowest wake, indicating that the width of the cold wake depends on the TC size.

The SST cooling reaches a peak during the day following a TC passage and the surface cold wake restores to normal in an e-folding time of 5–20 days [9, 10]. In the WNP, Mrvaljevic et al. [11] found that a cold wake induced by Typhoon Fanapi (2010) extended to more than 80 m depth and four days later a thin and warm surface layer was formed above the cold wake. The capped wake (defined as the layer of 26–27°C) returned to normal after an e-folding time of 23 days, almost twice that of the corresponding SST cooling. Similar phenomenon has also been observed in the WNP by D'Asaro et al. [7] and their observations showed that the subsurface wake commonly occurred after the passage of several typhoons.

2.2. In the Kuroshio region

As a famous western boundary current originating from the North Equatorial Current, the Kuroshio has unique properties, characterized by high temperature, high salinity, and large velocity. In the Kuroshio region, the mixed layer as well as the thermocline is usually deeper than that in the neighboring ocean water [12, 13], which restrains the typhoon-induced SST cooling. Based on satellite remote sensing data, Wu et al. [12] found that Typhoon Nari (2001) did not induce a significant SST cooling near the Kuroshio axis where the thermocline depth is 80–100 m, while a significant SST cooling occurred in the shelf region north of the Kuroshio where the thermocline depth is 20–30 m. In addition to the deep thermocline, they suggested that the strong advection of heat along the Kuroshio also contributed to weak SST cooling near the Kuroshio axis. Analyzing the SST responses to 22 typhoons which went across the Kuroshio in the East China Sea (ECS) from 2001 to 2010, Liu and Wei [13] demonstrated that the SST cooling (averaged within 150 km on the right of each typhoon track) in the Kuroshio region ranged from 0.61°C to 4.93°C with a mean of 2.09°C while the mean SST cooling in the neighboring ocean region was 2.68°C. Wei et al. [14] reported that a SST cooling of about 3°C in the Kuroshio was produced by Typhoon Megi (2004). Their results indicated that vertical mixing is mainly responsible for the SST cooling and the kinetic energy drawn from the Kuroshio baroclinic potential energy may contribute to the cooling by enhancing local vertical mixing.

There are two special cases of temperature cooling associated with typhoons in the Kuroshio region. First, a significant temperature drop of 4°C in the upper layer from the sea surface to about 100 m was observed in the Kuroshio region, near the southeast tip of Taiwan, before Typhoon Morakot (2009) passed over [15]. This was caused by an offshore cool jet which was generated by persistent westerly wind and stretched along the Kuroshio. Secondly, typhoon-induced inshore transport of Kuroshio subsurface water in the ECS can produce upwelling and cause severe drop in the SST on the shelf beside the Kuroshio. This happened off northeastern Taiwan Island during Typhoons Gerald (1987) and Hai-Tang (2005) [16–18].

Using 10-year satellite remote sensing SST data and Argo temperature profiles, Liu and Wei [13] investigated the temperature change (warming) in the surface and subsurface of Kuroshio when the temperature recovered from typhoon-induced temperature cooling. They found that the surface temperature change (1.24°C) in the Kuroshio region is slightly smaller than that (1.39°C) in the general ocean while the subsurface temperature change (3.52°C) in the former is much larger than that (1.52°C) in the latter. Their numerical simulations indicated that the subsurface temperature change is mostly caused by downwelling related to typhoon-induced

Ekman pumping. The warm water is extracted into the subsurface from the surface by the downwelling and subsequently moves downstream with Kuroshio current.

2.3. In the SCS

The SCS is the largest semi-enclosed marginal sea of the WNP. Many TCs pass through the SCS and some are born in this sea every year [3]. Compared with the open ocean of the WNP, the SCS displays a similar but some different temperature response to typhoons due to its unique hydrological environment and complex topography.

Using the Princeton Ocean Model, Chu et al. [19] showed that Typhoon Ernie (1996) induced a significant SST cooling with a rightward bias in the SCS, similar to that in the open ocean. But it also caused some unique responses such as the SST warming in the region from southwest of Taiwan Island to northwest of Luzon Island because of the convergence between the northward coastal current west of Luzon and the Kuroshio intrusion current through the Luzon Strait.

Owing to the shallow pre-typhoon mixed layer and thermocline, Typhoons Kai-Tak (2000), Lingling (2001) and Megi (2010) generated a very large SST drop of 10.8°C, 11°C and 8°C, respectively, in the SCS [8, 20, 21]. Chiang et al. [8] suggested that upwelling (account for 62%) dominated vertical mixing (31%) in producing the SST cooling under the influence of Kai-Tak, a weak and slowly moving typhoon. Megi's translation speed was 5.5–6.9 m/s over the ocean east of the Philippines, faster than 1.4–2.8 m/s over the SCS, and the pre-typhoon mixed layer depth in these two regions was about 40 m and 20 m, respectively [22]. As a result, the SST cooling in the former was only 1–2°C, quite smaller than that in the SCS. Based on the mooring observations in the northern SCS during Megi, Guan et al. [23] showed that the temperature cooling occurred in the entire observed water column (60–360 m), which was mainly caused by typhoon-induced upwelling.

After comparing the temperature responses to TCs in the SCS and in the tropical ocean of the NWP, Mei et al. [24] found that under the influence of TCs with an identical intensity, the SST cooling in the SCS is more than 1.5 times that in the tropical ocean, which could be attributed to the shallower mixed layer and stronger subsurface thermal stratification in the former. Numerical simulations showed that Typhoon Nuri (2008) induced a stronger SST cooling in the SCS than in the open ocean of the WNP when it travelled northwestward from the open ocean to the SCS [25]. Sun et al. [25] indicated that three processes are responsible for the different regional responses. Firstly, the SCS has a thinner mixed layer, which makes it easier to entrain cooler subsurface water into the surface layer. Secondly, the cyclonic background vorticity in SCS allows stronger current shears and turbulent eddy diffusivity to be generated, however, the background vorticity in the open ocean is anticyclonic. Finally, as the typhoon moved to higher latitude in SCS, the larger Coriolis frequency in the SCS is more favorable for producing stronger wind-current resonance and then stronger inertial amplitudes and turbulence.

http://dx.doi.org/10.5772/64009

3. Upper ocean current response to TCs

3.1. In the open ocean

Geisler [26] used a two-layer ocean model to investigate the linear response of ocean to a moving hurricane. He concluded that inertial-gravity waves are the dominant feature of the upper ocean response if TC's translation speed exceeds the phase speed of the first baroclinic mode, and inversely the oceanic response is a barotropic, geostrophical, and cyclonic gyre with upwelling in the storm's center. In the upper ocean, a TC typically generates near-inertial currents on the right side of the TC track (looking in the moving direction) in the Northern Hemisphere and causes the resonance between wind stress vectors and currents [4]. The wind stress rotates clockwise with time on the right of the track, which is homodromous with the near-inertial currents. In contrast, the wind stress rotates anticlockwise, against the near-inertial currents, on the left. The shear of near-inertial current across the mixed layer base can deepen the mixed layer. The near-inertial currents decay rapidly within a few inertial cycles, propagating downward into the thermocline and even deep ocean as near-inertial internal waves [27, 28].

Based on the observations from drifters during 1985–2009, Chang et al. [29] illustrated the composited near-surface ageostrophic currents under all recorded TCs with various intensity levels in the WNP. Strongest current is shifted to the right of the TC track. On average, the maximum velocity increases with the intensity of the TC and that for category 4 and 5 TCs may exceed 2.0 m/s. Moreover, they found that the near-surface current responses depend on the TC translation speed as presented by Geisler [26]. Hormann et al. [30] also observed a rightward shift of maximum near-inertial currents associated with Typhoon Fanapi (2010). Observed peak current magnitudes are up to 0.6 m/s and the e-folding decay time of the strong near-inertial currents within the cold wake is about 4 days.

Analyzing the observations of the upper ocean currents induced by four category-5 typhoons [Chaba (2004), Maon (2004), Saomai (2006) and Jangmi (2008)] in the WNP, Chang et al. [31] found that besides the rightward shift, the maximum mixed layer current velocity increased with the decreasing translation speed of the four typhoons. The maximum current velocities varied from 1.2 m/s to 2.6 m/s when the translation speed of typhoons changed from 8.1 m/s to 2.9 m/s. Additionally, the maximum current velocity shows a proportional relationship with the Saffir–Simpson hurricane scale of typhoons.

3.2. Influence on the Kuroshio current

The Kuroshio is a strong western boundary current in the WNP, similar to the Gulf Stream in the North Atlantic. Under the influence of Typhoon Hai-Tang (2005), the Kuroshio axis northeast of Taiwan Island moved onto the shelf [16, 18]. The average speed of the Kuroshio surface current increased by 18 cm/s after the typhoon's passage. The northward wind in the eastern part of the typhoon caused a coastal upwelling along the east coast of Taiwan Island and an east-west sea level slope was set up. This sea level slope generated a northward geostrophic current which enhanced the Kuroshio and pushed it onto the shelf northeast of Taiwan Island [18].

Zheng et al. [15] simulated and described the response of the Kuroshio to Typhoon Morakot (2009) in the region east of Taiwan Island. When Morakot came to the Kurosho, the southward wind before the typhoon center forced the surface flow of the Kuroshio to slow down to zero. Subsequently, the surface flow strengthened as Morakot got closer to the Kuroshio. There was a sudden speedup in the surface flow due to the disappearance of strong southward wind and the release of accumulated potential energy. Before Morakot approached the Kuroshio, the Kuroshio main stream was shifted eastward for more than 1.5°. The shifted main stream returned to its original location when the typhoon center went through the Kuroshio. As the surface flow slowed down, the Kuroshio main core shifted from the surface to the depth of 50–100 m and its maximum speed decreased from more than 1.3 m/s to less than 1.1 m/s.

Using three ADCPs deployed in the area east of Taiwan Island, Yang et al. [32] revealed that during the period of 2014–2015, the volume transport of the Kuroshio was reduced by six typhoons which moved almost from southeast to northwest in the region east of the island, but intensified by two typhoons which travelled northward.

3.3. In the SCS

As frequently appearing strong weather systems, typhoons have potential influence on local currents and large-scale circulations in the semi-enclosed SCS. The large-scale circulations in the SCS are controlled mainly by strong northeast monsoon wind in winter and by weak southwest monsoon wind in summer. As a result, a basin-wide cyclonic gyre appears in winter and it is replaced by a large diploe structure (a cyclonic gyre in the north and an anticyclonic gyre in the south) in summer. Under the influence of typhoons, both the cyclonic and anticyclonic gyres are intensified in summer while the northern and southern parts of the cyclonic gyre are intensified and weakened, respectively, in winter except October and November when both are intensified [33]. Additionally, accumulative effect of typhoons can affect mean mesoscale structures: weakening the cyclonic eddy northeast of Luzon Island and enhancing the cyclonic and anticyclonic eddies off Vietnam central coast [33].

Typhoons often trigger near-inertial waves or near-inertial oscillations (NIOs) in the SCS (e.g., [23, 34, 35]). Based on ADCP observations in the northern continental shelf of the SCS, Sun et al. [35] showed that Strong NIOs were generated by Typhoon Fengshen (2008) and lasted for about 15 days after the passage of the typhoon. A similar phenomenon was induced by Typhoon Chanchu (2004) in the west of the SCS [36]. Using the observations from an ADCP mooring deployed in the northern SCS, Chen et al. [34] found that Typhoon Nangka (2009) triggered an intensive NIOs while Typhoon Linfa (2009) did not, although they both passed by the mooring in the same month of June. This is because the mooring was located in the right of Nangka's moving track but in the left of Linfa's track. As a result, the wind stress affecting the mooring location rotated clockwise during Nangka but counterclockwise during Linfa.

Yang et al. [37] demonstrated that the second baroclinic mode dominated in the NIOs appearing after the passage of Typhoon Nesat (2011) in the northern SCS.

In the SCS, the signals of internal solitary waves (ISWs) often appear in current observations during the influence periods of typhoons. Xu et al. [38] observed a series of ISWs excited by a tropical storm Washi (2005) in the Northwestern SCS. The response of the ISWs was related to direct wind forcing and remote forcing from the inertial internal waves generated by Washi. Such ISWs were also observed in the northern SCS after the passage of Typhoon Neast (2011) [39] and in other marginal seas of the WNP after a typhoon passed over [40].

3.4. In the Taiwan Strait

A strait is a special sea area that connects two different sea waters. Thus, it is of great importance for material exchange and energy transfer between the two sea waters. Here we take the Taiwan Strait joining the ECS and the SCS as an example of strait in the WNP. The Taiwan Strait is a wide and long channel with an average depth of about 60 m, bounded by the Chinese Mainland to the west and Taiwan Island to the east.

Several typhoons pass over or pass by the Taiwan Strait every year. A typhoon can induce strong southward currents and reduce or reverse northward transport through the Taiwan Strait temporarily [41–44]. Chen et al. [41] observed that the northward currents in the northern end of the strait were reversed after the passage of Typhoons Rusa and Sinlaku in 2002. Based on buoy observations and numerical model simulations, Zhang et al. [42] found that five typhoons reversed northeastward current in the middle of the Taiwan Strait and induced five southward transport events through the strait during the period of 27 August to 5 October 2005. These southward transport events were directly forced by wind stress and/or along-strait water level gradient associated with the typhoons [42]. A similar southward transport event during Typhoon Krosa (2007) was simulated by Lin et al. [44]. The observations of the drifters deployed in the Taiwan Strait revealed current reversal when Typhoon Hai-Tang (2005) traversed the strait [43].

However, the southward transport event does not always occur under the effect of typhoons. Zhang et al. [45] identified four typhoons in 2005–2009 that enhanced northward transport through the Taiwan Strait. These typhoons travelled westward in the area south of the strait or moved northward from the south to the north along special tracks, resulting in a weak southward atmospheric forcing in the early stage and a strong northward atmospheric forcing in the later stage. Meanwhile, the effect of ageostrophic process generated by the atmospheric forcing also contributed to the enhanced northward transport.

The accumulative effect of all typhoons can modify monthly mean transports during the typhoon season and even annual mean transport through the Taiwan Strait. Based on numerically simulated results, Zhang et al. [46] demonstrated that if the effects of typhoons are considered, the monthly mean transport and annual mean transport are reduced by up to 0.45 Sv and 0.09 Sv (more than 10%), respectively, compared with those without typhoons.

4. Influence on mesoscale eddies

A mesoscale eddy is one vortex with its core surrounded by closed circulations. In the ocean, it has a time scale of 10–100 days and a dimension of 10–100 km in horizontal and 100–1000 m in vertical. It can travel a distance of 100–1000 km during its lifetime of several months. According to the rotation direction of closed circulations, mesoscale eddies are classified into cyclonic (counterclockwise) and anticyclonic (clockwise) eddies. Because the temperature inside a cyclonic eddy is usually lower than that of surrounding water, the cyclonic eddy is also called cold eddy. Inversely, an anticyclonic eddy is also called warm eddy. Sea level anomaly with respect to local mean sea level is negative in a cyclonic eddy whereas it is positive in an anticyclonic eddy.

Mesoscale eddies are ubiquitous in the open ocean and marginal seas (e.g., the SCS) of the WNP. There are two eddy-rich regions in the open ocean of the WNP [47–49]: the Kuroshio Extension region (30°–40°N, 140°E–180°W) and the North Pacific Subtropical Countercurrent region (18°–25°N, 122°E–160°E). The SCS is a well-known eddy-rich marginal sea with a large area of 3,500,000 km^2 and an average depth of over 2000 m [50, 51]. In these regions, typhoons appear frequently, which may influence the structure and evolution of some eddies.

Pre-existing cyclonic eddies can be intensified after a typhoon passes over. Zheng et al. [52, 53] revealed that the SST in pre-existing cyclonic eddies markedly decreased after the passage of Typhoon Hai-Tang (2005) because cold deep water was easily brought up due to uplifted thermocline, large current shear just below the mixed layer, and small thermal inertia within the cyclonic eddies. Chiang et al. [8] showed that when Typhoon Kai-Tak (2000) passed over the West Luzon Eddy (a cyclonic eddy), it produced an unusually intense SST drop of about 10.8°C and high chlorophyll *a* (Chl-*a*) concentration in the eddy. Lai et al. [54] found that a cold eddy was intensified when Typhoon Morakot (2009) passed by the eddy, resulting in a significant cooling. Nam et al. [55] pointed out that during Typhoon Man-Yi (2007), more distinct cooling of the SST and deepening of the mixed layer occurred within a cyclonic eddy with a thin mixed layer than within an anticyclonic eddy with a thick mixed layer. Sun et al. [56] demonstrated that four typhoons in 2004 successively enhanced and enlarged a cyclonic eddy in the Kuroshio meandering region, south of Japan, due to strong air-sea interaction and typhoon-induced upwelling. Although 49 super typhoons passed over 192 cyclonic eddies in the WNP during the period of 2000–2008, only about 10% of these eddies were intensified by the typhoons [57]. Thus, Sun et al. [57] concluded that the impact of typhoons on the strength of cyclonic eddies is inefficient.

Aside from the impact on the intensity of cyclonic eddies, typhoons can directly generate cyclonic eddies via strong wind stress curls and long-time forcing. Hu and Kawamura [58] found that three TCs separately generated a cyclonic eddy in the northern SCS where they had a loop track and provided sufficient forcing time for the eddy generation. Yang et al. [59] also reported that two mesoscale cyclonic eddies were separately induced in the SCS and the WNP by long time forcing of strong wind stress curls associated with Typhoons Hagibis and Mitag in 2007, respectively. Similar phenomenon has been revealed in other studies (e.g., [57, 60]). It is interesting that these cases happened in the SCS or in the west of the WNP where typhoons often travel slowly and easily change their moving direction. The main reason may be that this area is close to or located at the western edge of the North Pacific Subtropical High around which clockwise atmospheric circulation steers the movement of typhoons in the WNP to a great degree.

Compared with cyclonic eddies, the mixed layer is usually thicker and the thermocline is deeper within anticyclonic eddies. As a result, anticyclonic eddies are not significantly impacted by typhoons. Lin et al. [49] demonstrated that Super-typhoon Maemi (2003) caused a weak sea surface cooling of only about 0.5°C within an anticyclonic eddy, quite smaller than that (2°C) outside the eddy, while the typhoon was rapidly intensified due to the presence of the warm eddy. They reckoned that a thick mixed layer in the anticyclonic eddy prevented deep, cold water from being entrained into the surface layer and at the same time the warm water in the thick layer contributed to the intensification of the typhoon and sustained the typhoon's intensity due to a reduced negative feedback from typhoon-induced weak sea surface cooling. The results obtained by Nam et al. [55] showed that under the influence of Typhoon Man-Yi (2007), the entrainment within an anticyclonic eddy was weak because of the thick mixed layer, resulting in small changes of temperature profiles.

5. Storm surges and large waves

A typhoon usually induces low frequency abnormal variations in sea level (namely storm surges) and high frequency large waves (commonly called typhoon waves or storm waves). The main disaster causes accompanying with typhoons encompass storm surge, huge wave, strong wind and heavy rainfall. In low-lying coastal areas and islands, storm surge together with high spring tide is most destructive.

In the open ocean, the abnormal sea level is mainly controlled by pressure drop in the center of a TC according to the inverted atmospheric pressure effect. Although the cyclonic wind tends to reduce sea level height by Ekman transport, a round water bulge is often produced in response to low pressure in the center of the cyclone. At the same time, some long waves are induced and propagate forward faster than the cyclone itself. When these free waves arrive at the shallow and wide shelf waters near Mainland China, their wave height rises because of shallow depth and then their energy is gradually dissipated by bottom friction.

When a TC comes close to the coastal area, strong wind forcing is dominant in the generation of storm surges. The combination of onshore wind and low pressure results in a maximum storm surge near the TC center. The system of topography, bathymetry and coastline plays an important role in the distribution of storm surges. The inertial oscillations in storm surges generated by a strong typhoon would disappear quickly due to overdamping in the ultra-shallow water of the Bohai Sea. Surge waves can be induced by a typhoon and propagate counterclockwise or from north to south along the coasts in the China Seas, including the Bohai Sea, Yellow Sea, ECS, and SCS [61, 62]. Edge waves may appear when a typhoon moves in parallel with the coastline. The dissipation effect of bottom friction on these coastal trapped waves is significant in the wide shelfs of the China Seas. After a typhoon makes landfall and moves away, Ekman setup can generate surges along the coasts such as Tottori coasts of the East Sea [63].

Tides affect storm surges via tide-surge interaction in the shallow waters, which is significant in the coastal areas in the WNP [64–67]. In the Taiwan Strait, the tide-surge interaction is intensified because bottom friction is enhanced by strong tidal currents and storm-induced currents in the along-strait direction [68]. As a result of the tide-surge interaction, obvious oscillations appear in storm surges. The waves accompanying with a typhoon modulate storm surges by wave radiation stress, wave-dependent bottom shear stress and surface wind stress [69, 70]. Wave setup due to the breaking of wind waves associated with a typhoon can directly contribute to storm surges in the nearshore.

Typhoon-induced storm surges and huge waves often lead to loss of lives and damage to property in coastal regions and islands with a large population. Typhoon Saomai (2006) with a maximum wind of 75.8 m/s [71] and a lowest pressure of 915 hPa was the strongest one of typhoons making landfall on Mainland China in the past 50 years. When it made landfall at the coast of Zhejiang Province, the central pressure was still 920 hPa and a momentary maximum wind of 68 m/s was recorded. During its influence period, the highest wave recorded by a buoy (27.5°N,122.53°E) in the ECS was 8.6 m. Saomai induced devastating storm surges and a maximum storm surge of 4.01 m was observed at Aojiang station, Zhejiang Province. As reported in the Bulletin of Marine Disaster of China, the storm surges superposing on a high spring tide caused 230 persons dead and 96 missing, and a damage of more than 7 billion yuan to property. Another devastating typhoon, Haiyan, hit Philippines islands on 8 November 2013. Before its landfall, the maximum sustained wind was nearly 88 m/s and the momentary maximum wind was about 105 m/s. Storm surges induced by Typhoon Haiyan inundated a large coastal area in Philippines and caused a catastrophic damage [72, 73]. The inundation depth was up to more than 7 m in Tacloban city. During the typhoon, 6300 persons were killed and 1061 were still missing [72].

Storm surges, together with tides, mainly determine extreme sea levels and their spatial pattern along the coasts of the WNP [66, 74]. For some places where tidal range is small and tide-surge interaction is weak, storm surges play a more important role in the extreme sea levels [75]. Storm surges also contribute to the local increasing trend of annual maximum water level at coastal stations [76].

6. Biogeochemical and biological responses

TCs definitely have significant impacts on biogeochemical and biological processes in the oceans. Biological responses, including associated biogeochemical aspects, to typhoons have been observed in the WNP by ship surveys, buoys and satellite remote sensing data (e.g., [77–79, 20]), which will be discussed in the following text. Although typhoons may have direct and indirect effects on the carbon cycling in the WNP (e.g., [80–82]) and potentially contribute to global climate change, detailed contents of these effects are not included here. Many investigations show that various typical waters (open ocean waters, marginal seas, shelf waters) display some different biological responses to typhoons with unique mechanisms.

6.1. In open ocean water

In the deep ocean water of the WNP, Merritt-Takeuchi and Chiao [83] found an obvious growth of biological substances after the passage of Typhoon Xangsane (2006) which brought nutrients from the depths to the surface layer. There is a negative correlation between SST and Chl-*a* with the correlation coefficient of –0.67. Salyuk et al. [84] demonstrated that 81% of 123 TCs increased Chl-*a* concentration after the TC passage, which could last about 2 weeks. Based on multiple satellite observations and numerical experiments, however, Lin [85] showed that only two cases (18%) among 11 typhoons in 2003 induced phytoplankton blooms in the WNP.

The biological response of the upper ocean depends on the translation speed, spatial size, moving track and intensity of a typhoon and pre-existing ocean conditions. There is a positive correlation between Chl-*a* concentration and wind speed [83, 86]. Lin [85] found that 9 of 11 typhoons in 2003 did not induce phytoplankton blooms in the WNP. Among the 9 typhoons, eight typhoons had relatively small size, fast translation speed and insufficient wind intensity, and then they only caused weak responses in the ocean with a deep nutricline/mixed layer. Owing to the presence of warm eddy, the other typhoon, Maemi, was not able to induce phytoplankton bloom although it was very strong with a maximum wind of about 77 m/s. Typhoon Haitang (2005) did not enhance Chl-*a* in the sea area east of Taiwan because of high translation and a shallower MLD than the nutricline [87].

6.2. In the SCS

Typhoons are very active in the SCS. They often trigger phytoplankton blooms in this oligotrophic sea. Sun et al. [86] found that Typhoon Hagibis (2007) with a steep turn track had a significant effect on the surface Chl-*a* concentration in the SCS. Its long forcing time is favorable for the enhancement of Chl-*a* concentration although it is just a category 1 typhoon. Typhoon Kai-Tak (2000) increased surface Chl-*a* concentration by 30-fold on average in the SCS during three days [49]. As a result, Kai-Tak alone contributed 2–4% of the SCS's annual new primary production. A similar enhancement of Chl-*a* concentration appeared separately after typhoons Lingling (2001) and Damrey (2005) in the SCS [20, 88]. Zhao et al. [89] conservatively estimated that typhoons accounted for 3.5% of annual primary production in the SCS during the period of 1945–2005. Chen et al. [90] estimated that 5–15% of annual new primary production in the SCS was attributed to typhoons during 2003–2012.

Eddies are ubiquitous in the SCS. Some investigations have shown that Chl-*a* concentration enhanced by typhoons may be associated with cold core eddies in this area (e.g., [22, 91]). Chen and Tang [92] found that in the SCS a cold eddy was generated where Typhoon Linfa (2009) hovered, and subsequently an eddy-feature phytoplankton bloom was induced. The consistence between the bloom pattern and the cold eddy suggested that the typhoon-induced eddy potentially brought nutrients upward to the surface water, which contributed to the bloom.

Chen et al. [90] compared typhoon-enhanced primary production in the SCS with that in the subtropical ocean water of the WNP during the period of 2003–2012. Their results showed that the annual mean carbon fixation induced by typhoons was more in the SCS than in the ocean water. This is because the mixed layer is thicker and the nutricline depth is deeper in the latter in spite of its larger area and more super typhoons appearing there.

The biological response to typhoons can happen not only in the surface layer but also in the subsurface. Ye et al. [93] found that a Chl-*a* bloom appeared in the subsurface layer (20–100 m depth) of the SCS after the passage of Typhoon Nuri (2008) and lasted for three weeks. This subsurface bloom was stronger and its life was longer than the synchronous surface Chl-*a* bloom. Previous estimates of the contribution from typhoons to annual primary production in the SCS were mostly based on the results in the surface layer using remote sensing data. On this aspect, these estimates probably underestimate the actual contribution of typhoons.

6.3. In continental shelf waters

There are many wide and shallow continental shelf regions in the WNP, mostly located in the China Seas. The ecosystem in these shelf regions can become more productive after typhoons pass through [77, 78, 87].

The continental shelf of the southern ECS, northwest of Taiwan, is a typical oligotrophic and strong stratification area during summer. However, Shiah et al. [78] found that all chemical and biological parameters measured in a survey after the passage of Typhoon Herb (1996) were much larger than normal summer conditions. The typhoon caused primary production, particulate organic carbon (POC) concentration, bacterial production, and biomass to increase by at least two-fold. Their results indicated that wind mixing, re-suspension and terrestrial runoff associated with the typhoon were responsible for these responses. Based on multi-satellite observations, Chang et al. [16] showed that Typhoon Hai-Tang (2005) induced upwelling and increased Chl-*a* concentration in this shelf region, persisting for more than 10 days. The upwelling was likely caused by Ekman pumping due to strong typhoon wind and wind-driven shoreward intrusion of Kuroshio water along the shelf break. Siswanto et al. [94] also demonstrated that long-lasting southerly winds accompanying with typhoons can force Kuroshio current axis to move toward the shelf of the southern ECS, inducing upwelling. This process uplifts nutrients and then increases new productivity, which contributes 0.6–11.8% of the summer-fall new productivity in the ECS. In addition to the effects on primary production represented by Chl-*a*, typhoons may have influence on phytoplankton composition. Chung et al. [95] observed that a diatom bloom was induced in the southern ECS after the passage of Typhoon Morakot (2009) and the species composition was changed also.

Zhang et al. [96] observed that the surface Chl-*a* concentration in the continental shelf southeast of Hainan Island was increased by 38.5% after the passage of tropical storm Washi (2005). Chen et al. [97] showed that the primary productivity and nitrate-uptake-based new production in the upstream Kuroshio close to southern Taiwan were enhanced after the passage of three typhoons in 2007 by riverine mixing associated with the typhoons. After Typhoon Malou (2010) passed over Sagami Bay in the central part of Japan, both Chl-*a* concentration and bacterial

http://dx.doi.org/10.5772/64009

abundance increased at an inshore station due to terrestrial runoff and sediment resuspension while only Chl-*a* concentration rose at an offshore station due to terrestrial runoff [98].

There are some contrary cases. Zhou et al. [99] found that Typhoon Fengshen (2008) destroyed the pre-existing upwelling and meanwhile caused freshwater plume to spread in the continental shelf of the northeastern SCS, which led to nutrient-limited conditions. As a result, a negative phytoplankton growth rate appeared a week after the typhoon passage. Zhao et al. [100] also found that a sharp decrease of Chl-*a* concentration was caused by Typhoon Matsa (2005) in the nearshore area of the ECS.

Therefore, biological response to typhoons in the continental shelfs, especially nearshore areas, is very complex. It changes with different sea areas and different time periods, depending on specific circumstances such as geographical configuration, hydrological environment as well as the existence of rivers or not.

7. Summary

TCs usually change temperature and salinity vertical profiles by vertical mixing, upwelling and heat flux. Generally, the vertical mixing causes surface temperature cooling and subsurface warming. The upwelling brings cold deep water upward and makes the water temperature in the upper layer to decrease, which suppresses the subsurface warming and enhances the surface cooling. Surface sea water loses heat into air by heat flux and its temperature further decreases.

TCs typically generate energetic transient currents in the upper ocean, which modulates mean circulations such as the Kuroshio. The upper layer flow of the Kuroshio can be slowed or shut down temporarily just before a TC approaches. It can shift the location of the Kuroshio main stream. In the SCS, TCs affect both large scale and mesoscale circulations, and generate NIOs and ISWs. They change the instantaneous volume transport through a strait like the Taiwan Strait and then accumulatively modify its seasonal and annual mean transports. TCs not only influence pre-existing oceanic mesoscale eddies but also generate cyclonic eddies when they travel slowly.

TCs produce storm surges and huge waves which can flood coastal low-lying areas and threaten coastal structures. Because of wide continental shelves and flexural coastlines, the combination of onshore wind and low pressure accompanying with typhoons easily causes devastating storm surges along the coasts of the continent and islands in the WNP. Additionally, Ekman setup and wave setup also contribute to the storm surges in some shallow waters. In the continental shelf regions of the China Seas, coastal trapped waves are often induced and tide-surge interaction is significant in the shallow coastal sea areas and in the Taiwan Strait.

As the most intense case of air-sea interaction, TCs can cause biogeochemical and biological responses in the ocean. The vertical mixing and upwelling associated with TCs transport nutrients from the depths to the surface layer. As a result, phytoplankton blooms are often triggered and contribute to the primary productivity in the WNP, especially in the SCS. In continental shelf waters, the biological responses are more complex because of pre-existing upwelling, TC-enhanced freshwater plume, riverine mixing, terrestrial runoff, and sediment resuspension.

Since we focus on the influence of TCs in the WNP, some important aspects are not included here, such as the feedback of ocean to TCs. The accumulative effects of TCs on climate are not considered, either, because they are global and indirect, not limited to the WNP, in a long time scale. Regarding the complexity and extensiveness of ocean responses to TCs, some questions remain open and more observations and investigations are necessary to explore their answers in the future.

Acknowledgements

This work was jointly funded by the National Natural Science Foundation of China (Grants 41276007, U1305231 and 41076002) and the President Research Award (2013121047).

Author details

Wen-Zhou Zhang*, Sheng Lin and Xue-Min Jiang

*Address all correspondence to: zwenzhou@xmu.edu.cn

Department of Physical Oceanography, College of Ocean and Earth Sciences, Xiamen University, Xiamen, China

References

[1] Gray W M. A global view of the origin of tropical disturbances and storms. Monthly Weather Review. 1968;96:669-700

[2] Matsuura T, Yumoto M, Iizuka S. A mechanism of interdecadal variability of tropical cyclone activity over the western North Pacific. Climate Dynamic. 2003;21:105–117

[3] Wang G, Su J, Ding Y, Chen D. Tropical cyclone genesis over the South China Sea. Journal of Marine Systems. 2007;68(3):318–326

[4] Price J F. Upper ocean response to a hurricane. Journal of Physical Oceanography. 1981;11(2):153–175

[5] Lin I-I, Wu C-C, Pun I-F, Ko D-S. Upper-ocean thermal structure and the western North Pacific category 5 typhoons. Part I: ocean features and the category 5 typhoons' intensification. Monthly Weather Review. 2008;136(9):3288–3306

[6] Price J F, Sanford T B, Forristall G Z. Forced stage response to a moving hurricane. Journal of Physical Oceanography. 1994;24(2):233–260

[7] D'Asaro E A, Black P G, Centurioni L R, Chang Y-T, Chen S-S, Foster R C, Graber H C, Harr P, Hormann V, Lien R-C, Lin, I-I, Sanford T B, Tang T-Y, Wu C-C. Impact of Typhoons on the Ocean in the Pacific. Bulletin of the American Meteorological Society. 2014;95(9):1405–1418

[8] Chiang T-L, Wu C-R, Oey L-Y. Typhoon Kai-Tak: an Ocean's Perfect Storm. Journal of Physical Oceanography. 2011; 41(1): 221–233. DOI: 10.1175/2010jpo4518.1

[9] Mei W, Primeau F, McWilliams J C, Pasquero C. Sea surface height evidence for long-term warming effects of tropical cyclones on the ocean. Proceedings of the National Academy of Sciences. 2013;110(38):15207–15210

[10] Price J F, Morzel J, Niiler P P. Warming of SST in the cool wake of a moving hurricane. Journal of Geophysical Research: Oceans. 2008;113: C07010.

[11] Mrvaljevic R K, Black P G, Centurioni L R, Chang Y-T, D'Asaro E A, Jayne S R, Lee C M, Lien R-C, Lin I-I,Morzel J, Niiler P P, Rainville L, Sanford TB. Observations of the cold wake of Typhoon Fanapi (2010). Geophysical Research Letters. 2013;40(2):316–321

[12] Wu C-R, Chang Y-L, Oey L-Y, Chang C-W, Hsin Y C. Air-sea interaction between tropical cyclone Nari and Kuroshio. Geophysical Research Letters. 2008;35: L12605.

[13] Liu X, Wei J. Understanding surface and subsurface temperature changes induced by tropical cyclones in the Kuroshio. Ocean Dynamics. 2015;65(7):1017–1027

[14] Wei J, Liu X, Wang D-X. Dynamic and thermal responses of the Kuroshio to typhoon Megi (2004). Geophysical Research Letters. 2014;41(23):8495–8502

[15] Zheng Z-W, Zheng Q, Lee C-Y, Gopalakrishnan G. Transient modulation of Kuroshio upper layer flow by directly impinging typhoon Morakot in east of Taiwan in 2009. Journal of Geophysical Research: Oceans. 2014;119(7):4462–4473

[16] Chang Y, Liao H-T, Lee M-A, Chan J-W, Shieh W-J, Lee K-T, Wang G-H, Lan Y-C. Multisatellite observation on upwelling after the passage of Typhoon Hai-Tang in the southern East China Sea. Geophysical Research Letters. 2008; 35: L03612. DOI: 03610.01029/02007GL032858

[17] Tsai Y, Chern C-S, Wang J. Typhoon induced upper ocean cooling off northeastern Taiwan. Geophysical Research Letters. 2008; 35: L14605.

[18] Morimoto A, Kojima S, Jan S, Takahashi D. Movement of the Kuroshio axis to the northeast shelf of Taiwan during typhoon events. Estuarine, Coastal and Shelf Science. 2009;82(3):547–552

[19] Chu P-C, Veneziano J M, Fan C, Carron M J, Liu W-T. Response of the South China Sea to tropical cyclone Ernie 1996. Journal of Geophysical Research. 2000;105(C6): 13991–14009

[20] Shang S, Li L, Sun F, Wu J, Hu C, Chen D, Ning X, Qiu Y, Zhang C, Shang S. Changes of temperature and bio-optical properties in the South China Sea in response to Typhoon Lingling, 2001. Geophysical Research Letters. 2008; 35: L10602. DOI: 10.1029/2008GL033502

[21] Ko D-S, Chao S-Y, Wu C-C, Lin I-I. Impacts of typhoon Megi (2010) on the South China Sea. Journal of Geophysical Research: Oceans. 2014;119(7):4474–4489

[22] Chen X, Pan D, He X, Bai Y, Wang D. Upper ocean responses to category 5 typhoon Megi in the western north Pacific. Acta Oceanologica Sinica. 2012; 31(1): 51–58

[23] Guan S, Zhao W, Huthnance J, Tian J, Wang J. Observed upper ocean response to typhoon Megi (2010) in the Northern South China Sea. Journal of Geophysical Research: Oceans. 2014;119(5):3134–3157

[24] Mei W, Lien C-C, Lin I-I, Xie S-P. Tropical Cyclone–Induced Ocean Response: A Comparative Study of the South China Sea and Tropical Northwest Pacific. Journal of Climate. 2015;28(15):5952–5968

[25] Sun J, Oey L-Y, Chang R, Xu F, Huang S-M. Ocean response to typhoon Nuri (2008) in western Pacific and South China Sea. Ocean Dynamics. 2015;65(5):735–749

[26] Geisler J E. Linear theory of the response of a two layer ocean to a moving hurricane. Geophysical and Astrophysical Fluid Dynamics. 1970;1(1–2):249–272

[27] Price J F. Internal wave wake of a moving storm. Part I. Scales, energy budget and observations. Journal of Physical Oceanography. 1983;13(6):949–965

[28] Sanford, T B, Black P G, Haustein J R, Feeney J W, Forristall G Z, Price J F. ocean response to a hurricane. Part I: obersation. Journal of Physical Oceanography. 1987; 17: 2065–2083

[29] Chang Y-C, Chen G-Y, Tseng R-S, Centurioni L R, Chu P-C. Observed near-surface flows under all tropical cyclone intensity levels using drifters in the northwestern Pacific. Journal of Geophysical Research: Oceans. 2013;118(5):2367–2377

[30] Hormann V, Centurioni L R, Rainville L, Lee C-M, Braasch L J. Response of upper ocean currents to Typhoon Fanapi. Geophysical Research Letters. 2014;41(11):3995–4003

[31] Chang Y-C, Chu P-C, Centurioni L R, Tseng R-S. Observed near-surface currents under four super typhoons. Journal of Marine Systems. 2014;139:311–319

[32] Yang Y-J, Jan S, Chang M H, Wang J, Mensah V, Kuo T-H, Tsai C J, Lee C-Y, Andres M, Centurioni L R, Tseng Y-H, Liang W-D, Lai J-W. Mean structure and flctuations of the Kuroshio East of Taiwan from in situ and remote observations. Oceanography. 2015;28(4):74–83

[33] Wang G, Ling Z, Wang C. Influence of tropical cyclones on seasonal ocean circulation in the South China Sea. Journal of Geophysical Research: Oceans. 2009;114:C10022.

[34] Chen S, Hu J, Polton J A. Features of near-inertial motions observed on the northern South China Sea shelf during the passage of two typhoons. Acta Oceanologica Sinica. 2015;34(1):38–43

[35] Sun Z, Hu J, Zheng Q, Li C. Strong near-inertial oscillations in geostrophic shear in the northern South China Sea. Journal of oceanography. 2011b;67(4):377–384

[36] Sun L, Zheng Q, Wang D, Hu J, Tai C, Sun Z. A case study of near-inertial oscillation in the South China Sea using mooring observations and satellite altimeter data. Journal of Oceanography. 2011;67:677–687

[37] Yang B, Hou Y. Near-inertial waves in the wake of 2011 Typhoon Nesat in the northern South China Sea. Acta Oceanologica Sinica. 2014;33(11):102–111

[38] Xu Z-H, Yin B-S, Hou Y-J. Response of internal solitary waves to tropical storm Washi in the northwestern South China Sea. Annales Geophysicae. 2011; 29(11): 2181–2187. DOI: 10.5194/angeo-29-2181-2011

[39] Lin F, Hou Y, Liu Z, Hu P, Fang Y, Duan Y. The influence of background waves on internal solitary waves after the transit of Typhoon Neast in the northern South China Sea. Acta Oceanologica Sinica. 2014; 33(7): 40–47. DOI: 10.1007/s13131-014-0511-9

[40] Nam S, Kim D j, Kim H R, Kim Y G. Typhoon-induced, highly nonlinear internal solitary waves off the east coast of Korea. Geophysical Research Letters. 2007; 34(1). DOI: 10.1029/2006gl028187

[41] Chen T A, Liu C T, Chuang W S, Yang Y J, Shiah F K, Tang T Y, Chung S W. Enhanced buoyancy and hence upwelling of subsurface Kuroshio waters after a typhoon in the southern East China Sea. Journal of Marine Systems. 2003;42:65–79

[42] Zhang W Z, Hong H S, Shang S P, Yan X H, Chai F. Strong southward transport events due to typhoons in the Taiwan Strait. Journal of Geophysical Research: Oceans. 2009; 114: C11013.

[43] Chang Y C, Tseng R S, Centurioni L R. Typhoon-induced strong surface flows in the Taiwan Strait and Pacific. Journal of Oceanography. 2010;66(2):175–182

[44] Lin Y-H, Fang M-C, Hwung H-H. Transport reversal due to Typhoon Krosa in the Taiwan Strait. Open Ocean Engineering Journal. 2010;3:143–157

[45] Zhang W-Z, Hong H-S, Yan X-H. Typhoons enhancing northward transport through the Taiwan Strait. Continental Shelf Research. 2013;56:13–25

[46] Zhang W-Z, Chai F, Hong H-S, Xue H. Volume transport through the Taiwan Strait and the effect of synoptic events. Continental Shelf Research. 2014;88:117–125

[47] Yasuda I, Okuda K, Hirai M. Evolution of a Kuroshio warm-core ring variability of the hydrographic structure. Deep-Sea Research. 1992; 39: 131–161

[48] Qiu B. Seasonal eddy field modulation of the North Pacific Subtropical Countercurrent: TOPEX/Poseidon observations and theory. Journal of Physical Oceanography. 1999; 29: 1670–1685

[49] Lin I-I, Wu C-C, Emanuel K, Lee I-H, Wu C-R, Pun I. The interaction of Supertyphoon Maemi (2003) with a warm ocean eddy. Monthly Weather Review. 2005; 133: 2635–2649

[50] Wang G-H, Su J, Chu P-C. Mesoscale eddies in the South China Sea observed with altimeter data. Geophysical Research Letters. 2003; 30 (21): 2121. DOI: 10.1029/2003GL018532

[51] Xiu P, Chai F, Shi L, Xue H, Chao Y. A census of eddy activities in the South China Sea during 1993–2007. Journal of Geophysical Research: Oceans. 2010; 115: C03012. DOI: 10.1029/2009JC005657

[52] Zheng, Z W, Ho C R, Kuo N J. Importance of pre-existing oceanic conditions to upper ocean response induced by Super Typhoon Hai-Tang. Geophysical Research Letters. 2008; 35: L20603. DOI: 10.1029/2008GL035524

[53] Zheng Z-W, Ho C-R, Zheng Q, Lo Y T, Kuo N-J, Gopalakrishnan G. Effects of preexisting cyclonic eddies on upper ocean responses to Category 5 typhoons in the western North Pacific. Journal of Geophysical Research: Oceans. 2010; 115: C09013. DOI: 10.1029/2009JC005562

[54] Lai Q-Z, Wu L, Shie C l. Sea surface response temperature to typhoon marakot. Journal of Tropical Meteorology. 2015; 21(2): 111–120

[55] Nam S, Kim D, Moon W M. Observed impact of mesoscale circulation on oceanic response to Typhoon Man-Yi (2007). Ocean Dynamics. 2012; 62(1): 1–12. DOI: 10.1007/s10236-011-0490-8

[56] Sun L, Yang Y, Fu Y. Impacts of Typhoon on the Kuroshio Large Meander: Observation Evidences. Atmospheric and Oceanic Science Letters. 2009; 2(1): 45–50

[57] Sun L, Li Y-X, Yang Y-J, Wu Q, Chen X T, Li Q-Y, Li Y-B, Xian T. Effects of super typhoons on cyclonic ocean eddies in the western North Pacific: a satellite data-based evaluation between 2000 and 2008. Journal of Geophysical Research: Oceans. 2014; 119(9): 5585–5598. DOI: 10.1002/2013jc009575

[58] Hu J, Kawamura H. Detection of cyclonic eddy generated by looping tropical cyclone in the northern South China Sea a case study. Acta Oceanologica Sinica. 2004; 23(2): 213–224

[59] Yang, Y-J, Sun L, Duan A M, Li Y-B, Fu Y-F, Yan Y-F, Wang Z Q, Xian T. Impacts of the binary typhoons on upper ocean environments in November 2007. Journal of Applied Remote Sensing. 2012; 6(1): 063583–063581. DOI: 10.1117/1.jrs.6.063583

[60] Tseng Y-H, Jan S, Dietrich D E, Lin I-I, Chang Y-T, Tang T-Y. Modeled oceanic response and sea surface cooling to Typhoon Kai-Tak. Terrestrial Atmospheric and Oceanic Sciences. 2010; 21(1): 85–98. DOI: 10.3319/tao.2009.06.08.02(iwnop)

[61] Liu F, Wang X. A review of storm-surge research in China. Natural Hazards. 1989; 2:17–29

[62] Ding Y, Yu H, Bao X, Kuang L, Wang C, Wang W. Numerical study of the barotropic responses to a rapidly moving typhoon in the East China Sea. Ocean Dynamics. 2011; 61(9): 1237–1259. DOI: 10.1007/s10236-011-0436-1

[63] Kim S, Matsumi Y, Yasuda T, Mase H. Storm surges along the Tottori coasts following a typhoon. Ocean Engineering. 2014; 91: 133–145. DOI: 10.1016/j.oceaneng.2014.09.005

[64] Pan H, Liu F. A numerical study of the tide-surge interaction in the East China Sea and the South China Sea. Chinese Journal of Oceanology and Limnology. 1994; 12(1): 13–21

[65] Park Y H, Suh K-D. Variations of storm surge caused by shallow water depths and extreme tidal ranges. Ocean Engineering. 2012; 55: 44–51

[66] Feng X, Tsimplis M N. Sea level extremes at the coasts of China. Journal of Geophysical Research: Oceans. 2014; 119(3): 1593–1608. DOI: 10.1002/2013jc009607

[67] Xu J-L, Zhang Y-H, Cao A-Z, Liu Q, Lv X-Q. Effects of tide-surge interactions on storm surges along the coast of the Bohai Sea, Yellow Sea, and East China Sea. Science China Earth Sciences. 2015; 9: 1–9. DOI: 10.1007/s11430-015-5251-y

[68] Zhang W-Z, Shi F, Hong H-S, Shang S-P, Kirby J T. Tide-surge Interaction Intensified by the Taiwan Strait. Journal of Geophysical Research: Oceans. 2010; 115: C06012. DOI: 10.1029/2009jc005762

[69] Zhang M-Y, Li Y-S. The synchronous coupling of a third-generation wave model and a two-dimensional storm surge model. Ocean Engineering. 1996; 23(6): 533–543

[70] Yin B-S, Xu Z-H, Huang Y, and Lin X. Simulating a typhoon storm surge in the East Sea of China using a coupled model. Progress in Natural Science. 2009; 19: 65–71

[71] Liu D, Pang L, Xie B. Typhoon disaster in China: prediction, prevention, and mitigation. Natural Hazards. 2009; 49(3): 421–436. DOI: 10.1007/s11069-008-9262-2

[72] Lagmay, A. M. F., Agaton, R. P., Bahala, M. A. C., Briones, J. B. L. T., Cabacaba, K. M. C., Caro, C. V.C., Dasallas, L. L., Gonzalo, L A L., Ladiero, C. N., Lapidez, J. P., Mungcal, M. T. F., Puno, J. V. R., Ramos, M. M. A. C., Santiago, J., Suarez, J. K., Tablazon, J. P. Devastating storm surges of Typhoon Haiyan. International Journal of Disaster Risk Reduction. 2015; 11: 1–12. doi:10.1016/j.ijdrr.2014.10.006

[73] Shibayama, T. Field surveys of recent storm surge disasters. Procedia Engineering. 2015; 116: 179–186

[74] Zhang H, Sheng J-Y. Examination of extreme sea levels due to storm surges and tides over the northwest Pacific Ocean. Continental Shelf Research. 2015; 93: 81–97

[75] Feng J, Storch H, Jiang W, Weisse R. Assessing changes in extreme sea levels along the coast of China. Journal of Geophysical Research: Oceans. 2015; 120: 8039–8051. DOI: 10.1002/2015JC011336

[76] Feng J, Jiang W. Extreme water level analysis at three stations on the coast of the Northwestern Pacific Ocean. Ocean Dynamics. 2015; 65(11): 1383–1397. DOI: 10.1007/s10236-015-0881-3

[77] Chang J, Chung C-C, Gong G-C. Influences of cyclones on chlorophyll a concentration and Synechococcus abundance in a subtropical western Pacific coastal ecosystem. Marine Ecology Progress Series. 1996; 140: 199-205. DOI: 10.3354/meps140199

[78] Shiah F K, Chung S-W, Kao S-J, Gong G-C, liu K-K. Biological and hydrographical responses to tropical cyclones (typhoons) in the continental shelf of the Taiwan Strait. Continental Shelf Research. 2000; 20: 2029–2044

[79] Lin I-I, Liu W T, Wu C-C, Chiang J C H, Sui C-H. Satellite observations of modulation of surface winds by typhoon-induced upper ocean cooling. Geophysical Research Letters. 2003; 30(3): 1131. DOI: 10.1029/2002gl015674

[80] Nemoto K, Midorikawa T, Wada A, Ogawa K, Takatani S, Kimoto H, Ishii M, Inoue H Y. Continuous observations of atmospheric and oceanic CO_2 using a moored buoy in the East China Sea: variations during the passage of typhoons. Deep Sea Research Part II: Topical Studies in Oceanography. 2009; 56(8–10): 542–553. DOI: 10.1016/j.dsr2.2008.12.015

[81] Hung C-C, Gong G-C, Chou W-C, Chung C-C, Lee M-A, Chang Y, ... Laws E. The effect of typhoon on particulate organic carbon flux in the southern East China Sea. Biogeosciences Discussions. 2010; 7(3): 3521–3550. DOI: 10.5194/bgd-7-3521-2010

[82] Wada A, Midorikawa T, Ishii M, Motoi T. Carbon system changes in the East China Sea induced by Typhoons Tina and Winnie in 1997. Journal of Geophysical Research: Oceans. 2011; 116: C07014. DOI: 10.1029/2010jc006701

[83] Merritt-Takeuchi A M, Chiao S. Case Studies of Tropical Cyclones and Phytoplankton Blooms over Atlantic and Pacific Regions. Earth Interactions. 2013; 17(17): 1–19. DOI: 10.1175/2013ei000517.1

[84] Salyuk P A, Golik I A, Stepochkin I E. Satellite remote sensing using for analysing of chlorophyll – "a" concentration changes during tropical cyclones passing in northwestern Pacific. Asia-Pacific Journal of Marine Science & Education. 2014; 4(1): 111

[85] Lin I-I. Typhoon-induced phytoplankton blooms and primary productivity increase in the western North Pacific subtropical ocean. Journal of Geophysical Research: Oceans. 2012; 117: C03039. DOI: 10.1029/2011JC007626

[86] Sun L, Yang Y, Xian T, Lu Z, Fu Y. Strong enhancement of chlorophyll a concentration by a weak typhoon. Marine Ecology Progress Series. 2010; 404: 39–50. DOI: 10.3354/meps08477

[87] Shan H, Guan Y, Huang J. Investigating different bio-responses of the upper ocean to Typhoon Haitang using Argo and satellite data. Chinese Science Bulletin. 2014; 59(8): 785–794. DOI: 10.1007/s11434-013-0101-9

[88] Zheng G-M, Tang D-L. Offshore and nearshore chlorophyll increases induced by typhoon winds and subsequent terrestrial rainwater runoff. Marine Ecology Progress Series. 2007; 333: 61–74. DOI: 10.3354/meps333061

[89] Zhao H, Tang D, Wang Y. Comparison of phytoplankton blooms triggered by two typhoons with different intensities and translation speeds in the South China Sea. Marine Ecology Progress Series 2008; 365: 57-65

[90] Chen X, Pan D, Bai Y, He X, Arthur Chen C T, Kang Y, Tao B. Estimation of typhoon-enhanced primary production in the South China Sea: a comparison with the Western North Pacific. Continental Shelf Research. 2015; 111: 286–293. DOI: 10.1016/j.csr.2015.10.003

[91] Shang X, Zhu H, Chen G, Xu C, Yang Q. Research on cold core eddy change and phytoplankton bloom induced by typhoons: case studies in the South China Sea. Advances in Meteorology. 2015; 1–19. DOI: 10.1155/2015/340432

[92] Chen Y, Tang D. Eddy-feature phytoplankton bloom induced by a tropical cyclone in the South China Sea. International Journal of Remote Sensing. 2012; 33(23): 7444–7457. DOI: 10.1080/01431161.2012.685976

[93] Ye H J, Sui Y, Tang D-L, Afanasyev Y D. A subsurface chlorophyll a bloom induced by typhoon in the South China Sea. Journal of Marine Systems. 2013; 128: 138–145. DOI: 10.1016/j.jmarsys.2013.04.010

[94] Siswanto E, Morimoto A, Kojima S. Enhancement of phytoplankton primary productivity in the southern East China Sea following episodic typhoon passage. Geophysical Research Letters. 2009; 36: L11603. DOI: 10.1029/2009GL037883

[95] Chung C-C, Gong G-C, Hung C-C. Effect of Typhoon Morakot on microphytoplankton population dynamics in the subtropical Northwest Pacific. Marine Ecology Progress Series. 2012; 448: 39–49. DOI: 10.3354/meps09490

[96] Zhang S, Xie L, Hou Y, Zhao H, Qi Y, Yi X. Tropical storm-induced turbulent mixing and chlorophyll-a enhancement in the continental shelf southeast of Hainan Island. Journal of Marine Systems. 2014; 129: 405–414. DOI: 10.1016/j.jmarsys.2013.09.002

[97] Chen Y, Chen H-Y, Jan S, Tuo S-H. Phytoplankton productivity enhancement and assemblage change in the upstream Kuroshio after typhoons. Marine Ecology Progress Series. 2009; 385:111-126. DOI: 10.3354/meps08053

[98] Tsuchiya K, Kuwahara V S, Hamasaki K, Tada Y, Ichikawa T, Yoshiki T, Nakajima R, Imai A, Shimode S, Toda T. Typhoon-induced response of phytoplankton and bacteria in temperate coastal waters. Estuarine, Coastal and Shelf Science. 2015; 167: 458–465. DOI: 10.1016/j.ecss.2015.10.026

[99] Zhou L, Tan Y, Huang L, Huang J, Liu H, Lian X. Phytoplankton growth and microzooplankton grazing in the continental shelf area of northeastern South China Sea after Typhoon Fengshen. Continental Shelf Research. 2011; 31(16): 1663–1671. DOI: 10.1016/j.csr.2011.06.017

[100] Zhao H, Shao J, Han G, Yang D, Lv J. Influence of Typhoon Matsa on Phytoplankton Chlorophyll-a off East China. PLoS One. 2015; 10(9): 1–13. DOI: 10.1371/journal.pone.0137863

Chapter 2

Diagnosing Tropical Cyclone Rapid Intensification Through Rotated Principal Component Analysis of Synoptic-Scale Diagnostic Fields

Alexandria Grimes and Andrew E. Mercer

Additional information is available at the end of the chapter

http://dx.doi.org/10.5772/63988

Abstract

Forecasts of rapid intensification (RI) within tropical cyclones continue to be a major challenge, primarily due to difficulty in determining the processes that distinguish RI and non-RI storms. In this study, the aim was to identify the most important RI/non-RI discriminatory variables in the North Atlantic basin, not only by level, but also spatial location relative to the tropical cyclone center. These important variables, identified using rotated principal component analysis on one-dimensional and three-dimensional GEFS reforecast base-state variables from 1985 to 2009, led to the identification of diagnostic fields with the largest variability between RI and non-RI events. Hierarchical clustering techniques performed on rotated PC loadings provided map types of RI and non-RI cyclones. Analysis of these composite map types, as well as composite derived fields including divergence, relative vorticity, equivalent potential temperature, static stability, and vertical shear, revealed interesting distinguishing characteristics between RI and non-RI events. Results suggested that vorticity in the mid-levels, divergence in the upper-levels, equivalent potential temperature, and specific humidity play critical roles in successfully discriminating between RI and non-RI storms. These findings give key insights to which variables should be used in developing a prognostic classification scheme to assist with operational forecasts of tropical cyclone RI.

Keywords: tropical cyclone, rapid intensification, principal component analysis, cluster analysis, kinematic and thermodynamic tropical meteorology

1. Introduction

Modern statistical and dynamic forecast models continue to demonstrate low forecast skill in identifying the onset of rapid intensification (hereafter known as RI) within tropical cyclones (hereafter known as TCs). Even though storms which rapidly intensify can cost governments billions of dollars in damage upon landfall (e.g. by destroying property through flooding–as with Katrina in 2005) and RI forecasting is considered one of the top priorities for the National Hurricane Center [1], little advancement has been made in improvement of probabilistic tropical cyclone RI forecasting. While previous studies have examined the intensification patterns between RI and non-RI TCs [2, 3], the technological impairments, coupled with the complexity of these systems, have left gaps in understanding the large-scale structures associated with RI storms [1–4]. These gaps result in poor statistical forecast model accuracy, which requires prior knowledge of relevant RI variables. While recent research shows modest improvements in RI forecasts, and global models have steadily improved in their ability to predict the large-scale environmental conditions of TCs [3], forecast skill scores still remain inadequate.

Current statistical forecast models blend both thermodynamic and kinematic variables in attempts to increase the skill, emphasizing meteorological processes deemed more crucial to RI prediction [1, 4–6]. Improvements to the Statistical Hurricane Intensity Prediction Scheme Rapid Intensification Index (SHIPS-RII) continue to be added regularly since the original implementation by the National Hurricane Center (NHC) in 2004 for the North Atlantic [1]. The latest enhanced SHIPS-RII consists of 10 predictors, including previous 12-hour intensity change, vertical shear, divergence at 200 hPa, total precipitable water, GOES-IR imagery, potential intensity, oceanic heat content, max sustained wind, and an inner-core dry air predictor [1]. Despite the addition of new predictors, Brier skill scores (BSS) relative to climatology for Atlantic RI forecasts remain below 20% [1]. Additionally, verification of all operational consensus intensity forecast models for the NHC, including their official intensity forecast, showed only limited improvement as Peirce skill scores remained below 0.2 [1]. Other studies have included predictors that resolve the inner-core environment more effectively, utilizing microwave passive imagery predictors in a probabilistic logistic regression (LR) model. Despite this effort, BSS values only improved to roughly 22% with either simulated real-time LR models or LR models utilizing reanalysis data [4]. Additionally, using a baseline peak wind speed of 25 knot intensity (at all RI thresholds) severely reduces skill to below 15% when compared to a probabilistic LR model utilizing current SHIPS parameters previously developed in [7].

In order to improve statistical model prediction of the onset of TC RI, the ability to identify distinguishing meteorological characteristics of the storm structure between RI and non-RI TCs with 24 hours lead time is crucial. While research of this nature is not new [2, 3], the approaches have differed (e.g. data selection, data reduction, meteorological variables chosen, compositing approaches). For example, Kaplan and DeMaria [2] and Kaplan et al. [8] noted that RI was more likely to occur for TCs that were situated over regions of higher than average sea surface temperature (SST), strong upper-level divergence, large low- to mid-tropospheric moisture, and weaker than average vertical wind shear [2, 8]. Other research (see [3]) also observed that RI events occurred in environments with weaker deep-layer shear (as was found in [2, 8]) and greater conditional instability in the Atlantic basin than non-RI events. This research also noted that TCs moving over a warm ocean anomaly were found to be equally likely to intensify slowly or rapidly given other assumptions are met [3], a result in contrast with the work shown in [2, 8]. While recent research suggests environmental, internal dynamic processes, and oceanic conditions [2, 3, 8] all play a role in RI, research performed in [3] concluded RI is mostly controlled by internal dynamical processes, provided a pre-existing favorable environment exists. Research performed in [4] reiterated this sentiment suggesting Atlantic basin forecasts benefit more from the inclusion of storm structure information (more than the Pacific basins), which has yet to be explained.

In an effort to continue improving the understanding of the internal dynamics of TCs undergoing RI, the current study sought to identify important diagnostic variables in the North Atlantic basin, looking not only at which levels, but also at which spatial points in proximity to the cyclone are distinguishable between the two types of systems. The primary research question being considered is: What meteorological parameters discriminate RI from non-RI storms most effectively, and what spatial location in the TC domain provide the largest differences in these fields? These findings give key insights to which variables should be used in future development of a prognostic artificial intelligence classification scheme to assist with operational forecasts of RI. Section 2 provides a description of the data and methodology, while Section 3 presents the results of the work and Section 4 provides a discussion and conclusions.

2. Data and methodology

2.1. Dataset description

While the NHC defines RI as an increase in wind of 30 knots (kt) in 24 hours, several RI definitions are usually considered during the research phase of model development [1, 4]. This study examined three separate definitions of RI (following [1]), including the operational definition of a 30kt increase of wind speed in 24 hours and two experimental definitions of 25kt and 40kt increases. All Atlantic tropical and subtropical systems, from 1985–2009 from the NHC Atlantic best track data (HURDAT–[9]) were considered. For the three different RI definitions, the full database of 298 TC events were divided into RI and non-RI groups, yielding 152 RI and 146 non-RI cases with the 25kt definition, 119 RI and 179 non-RI for the 30kt definition, and 46 RI and 252 non-RI for the 40kt definition (**Figure 1** breaks these down by Saffir-Simpson scale category). Since a forecast proxy was desired, base-state meteorological fields from the National Centers for Environmental Prediction (NCEP) Global Ensemble Forecast System (GEFS–[10]) reforecast database were retained 24 hours prior to the period of greatest intensification for all storms (RI and non-RI). GEFS reforecast data are provided at a 1° resolution at 3-hour forecast intervals from 0 to 72-hours. Three-dimensional base-state meteorological fields at eight vertical levels (1000–100 hPa) were utilized, including: geopotential height, temperature, u and v wind components, and specific humidity. Additionally, single-layer variables were considered, including mean sea level pressure (MSLP), skin temperature (a proxy for SST), latent heat flux, sensible heat flux, convective available potential energy (CAPE), convective inhibition (CIN), and vertical velocity at 850 hPa were evaluated.

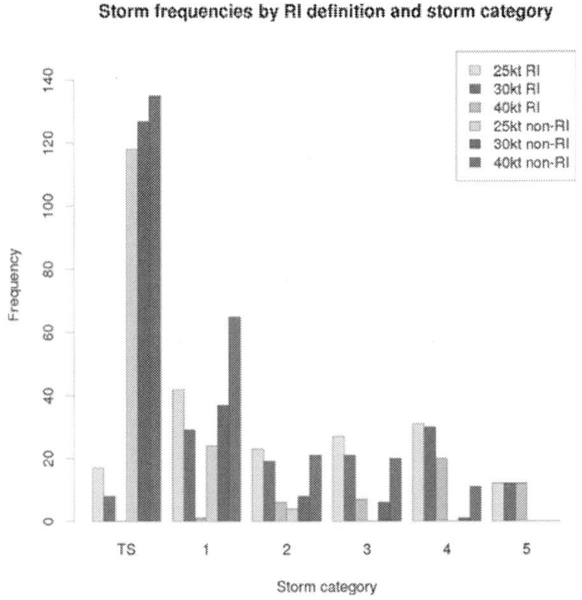

Figure 1. Distribution of category per TC event type per RI definition using 25kt/24-hours, 30kt/24-hours, and 40kt/24-hours definitions.

As a primary goal was to diagnose RI using TC structure relative to the storm center, storm-centric GEFS reforecast domains for each cyclone were obtained. Storm centers were identified by determining the local minimum in GEFS MSLP nearest the NHC-defined TC center, 24-hours prior to the timestep associated with the greatest intensification. Each variable was retained on a 15° × 11° latitude/longitude grid centered on this domain. In the event multiple occurrences of peak intensification occurred for an individual TC (which occurred 28 times when using 25kt/24-hours, 13 times for 30kt/24-hours, and once for 40kt/24-hours), the first was chosen. Thus, the results presented herein deal with the first instance of peak intensification regardless of the frequency of peak intensification for a given TC.

2.2. RPCA

As the primary goal of this research was the identification of variables and spatial locations most favorable for distinguishing RI and non-RI storms, discriminatory statistical methods were needed. One method, rotated principal component analysis (RPCA), has been shown to be useful in discriminating meteorological environments of different types [11–14]. These studies also used permutation testing to evaluate magnitude differences in diagnostic variables for each environment. Both of these techniques were utilized in the current study so that both spatial configuration and magnitude difference could be assessed.

2.2.1. S-mode RPCA

The first approach to RPCA, S-mode analysis [13], provided a diagnosis of the spatial relationship among gridpoints for all cases. For S-mode, the similarity matrix is computed on the individual spatial locations and is eigenanalyzed to identify particular locations that group together. The S-mode rotated principal component (RPC) loadings are maps that demonstrate these spatial relationships (known as modes of variability), with the RPC scores revealing the similarity between the individual cases and the resulting S-mode loading maps. To reduce the dimensions of the eigenvector matrix, truncation of RPCs was completed by evaluating a scree plot, as well as using a congruence test. A congruence test is a way to measure pattern and magnitude similarity of a dataset, corresponding to the cosine of the angular separation between the loadings, by maximizing the dissimilarity of the two loading patterns [15]. The congruence coefficient presenting a strong relationship for any absolute value greater than 0.81 was marked as the truncation point. RI and non-RI datasets (consisting only of base-state variables for all 298 cases) were combined, where the analysis of both RI and non-RI event

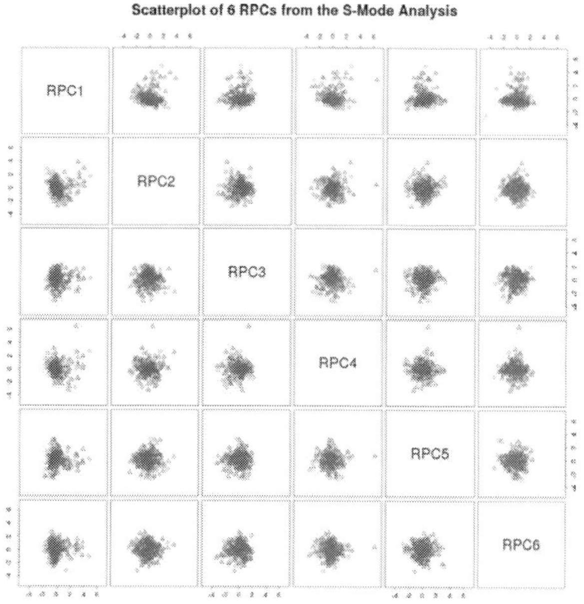

Figure 2. Pairwise scatterplots of all six PC score vectors. RI PC scores are redpoints, while non-RI points are blue. The significant overlap among the groups demonstrates the challenges in linearly separating these types of TCs.

deviations and the loading patterns provided information on how the systems are grouping together (e.g. cooler SSTs versus warmer, upper level trough/ridge patterns, and influence of land at the surface 24-hours prior). To demonstrate the lack of linear separability in the resulting RPCs, a pairwise scatterplot of all six PC score vectors was formulated (**Figure 2**). There is significant overlap among the RI and non-RI PC scores, rendering separation via classification very difficult, motivating the need to consider additional analysis techniques.

		Variance explained (%)					
		25kt/24-hours		30kt/24-hours		40kt/24-hours	
	S-mode	T-mode		T-mode		T-mode	
RPCs	Combined	RI	non-RI	RI	non-RI	RI	non-RI
1	24	18	13	14	12	16	11
2	12	11	12.8	11	13	11	14
3	7.4	7.6	4.8	7.7	5.0	7.3	6.3
4	3.6	6.6	7.7	7.2	7.8	7.0	8.4
5	4.5	4.7	4.7	4.9	5.8	5.4	4.8
6	3.3	—	4.5	—	—	—	—

Table 1. Rotated principal components variance explained for S-mode, as well as T-mode, for each event type and each RI definition. Dashes simply mean that this number of RPCs was not retained based on the previously described testing methodologies.

2.2.2. *T-mode RPCA*

While S-mode helped reveal the difficulties in identifying relevant RI/non-RI distinguishing characteristics, the results did not provide the necessary discrimination capability of interest in this work. Recent work has shown the value of composite analysis with T-mode RPCA in identifying discriminating characteristics for different meteorological event types [11, 12]. Following the methodology of [11, 12], a T-mode varimax-rotated RPCA [11, 16], conducted simultaneously on all GEFS reforecast fields, was completed on all RI events and all non-RI events separately. T-mode contrasts S-mode in that in T-mode, the relationships between events, as opposed to spatial locations, are of interest, and thus the correlation matrix is computed on the event dimension of the data. Following methods established in [11, 12], the resulting uncorrelated eigenvector matrix and associated eigenvalues reduced to a subset of RPCs for each event type and each RI definition (**Table 1**). Similar to the S-mode RPCA approach, the truncation point was determined through utilization of a scree plot and the congruence test. The resulting RPC loadings maintain the same dimension as the event dimension, so events were clustered by RPC loading magnitude using hierarchical clustering with Ward's minimum variance method [16]. To assess cluster quality, a cluster verification statistic (silhouette coefficient [17]) was found that includes two components:

1. a measure of intra-cluster spread (cluster cohesion–should be small) and

2. a measure of inter-cluster spread (cluster separation–should be large) [11].

In this study, the mean of the silhouette coefficient values for all events considered in the cluster analysis was retained as a measure of cluster analysis performance. With the silhouette coefficient, values approaching 1 suggest a minimization of cluster cohesion and a maximization of cluster separation. Negative values suggest a particular event was misclustered. The cluster analysis revealed six clusters each for RI and non-RI storm types using the 25kt/24-hours definition, seven non-RI and six RI using the 30kt/24-hours definition, and seven non-RI and five RI using the 40kt/24-hours definition (**Table 2** provides the number of events per cluster, as well as silhouette coefficient values). Events within each cluster were averaged together, yielding map types that retained unique synoptic-scale structures and provided more detailed map types of RI and non-RI TC environments than simply averaging all events together. The resulting composites allowed for the identification of spatial structure among RI/non-RI events.

	T- mode cluster distribution and Silhouette coefficients (S.C.)		
RI	**25kt/24-hours**	**30kt/24-hours**	**40kt/24-hours**
S.C.	0.24	0.24	0.26
1	22	18	9
2	28	21	**14**
3	18	25	8
4	13	27	9
5	**40**	**28**	6
6	31	–	–
non-RI			
S.C.	0.20	0.21	0.22
1	23	25	31
2	26	21	40
3	**37**	15	**55**
4	16	18	33
5	21	38	24
6	23	**47**	28
7	–	15	41

Table 2. Silhouette coefficients and number of events per cluster through Ward's method on T-mode RPC loadings.

2.3. Permutation testing

While the composites resulting from the RPCA approach are useful for diagnosing spatial characteristics within RI and non-RI environments, magnitude differences are diagnosed more effectively using hypothesis testing. In this study, permutation tests [16] comparing magnitudes of diagnostic fields in RI and non-RI storms were utilized at each gridpoint from the study domains, yielding a spatial map of significance values associated with each variable tested. The resulting plots provided specific regions in the study domain where statistically significant magnitude differences between RI and non-RI storms existed for individual GEFS reforecast variables. These results provided insight not only into the scope of these magnitude differences but into the spatial locations of the differences, which complement the RPCA results well.

3. Analysis

Based on previous work done [12], variables selected for evaluation consisted of base-state variables and derived variables as listed in Section 2.1. Through the S-mode technique, only base-state variables were examined; therefore, only notable characterizations are summarized in the text below. The T-mode and cluster analysis technique; however, yielded numerous composite fields for consideration. To minimize this impact, the cluster from each RI group and each RI definition that contained the largest number of events (bolded in **Table 2**) is provided for discussion below.

3.1. RI and non-RI S-mode maps

As stated previously, S-mode analysis of the base-state meteorological fields was conducted first. **Table 1** shows that RPC1 contained the largest variance explained (roughly 24%), and the loading map (**Figure 3**) revealed that areas of higher heights in the northeast quadrant quadrant and over the storm center co-varied with higher MSLP, lower temperatures, lower moisture and latent heat content. A total of 16 events' RPC scores (**Table 3**) exceeded 2 standard deviations above the mean, suggesting strong positive correlation between those events and RPC1. Of these 16 events, six were classified as RI cases with the 25kt/24-hour definition of RI. In fact, the highest positive deviation (approximately 5 standard deviations above the mean) was an RI case, while the second largest positive deviations were a mixture of RI and non-RI events. These results suggest a blend of RI and non-RI events for RPC1. Similarly, RPC2 results (which explained approximately 12% of the variability) revealed lower heights in proximity of the storm center (**Figure 4**) co-varied with lower MSLP, cooler low-level temperatures in the southwest quadrant of the cyclone and overall higher specific humidity and latent heat flux values. Additionally, patterns revealing a wrap-around of moisture over the storm-center were revealed. However, only three of the eight events that exceeded 2 standard deviations from the mean were RI events, again showing the blending of RI and non-RI cases in these results. RPC3 (which explained approximately 7% of the total variance), exhibited higher heights co-located with higher MSLP, temperature, specific humidity, and latent heat flux over the storm center and to the south, but lower heights, temperature, specific humidity, and latent heat flux North of the storm center (**Figure 5**). This RPC profile, along with RPCs 4–6 (not shown), is indicative of baroclinic environmental influence associated with both storm types. These first three RPCs, explaining over half of the variability combined, demonstrated a recurring problem in the S-mode analysis, namely the inability to separate RI and non-RI events (as was seen in **Figure 2**).

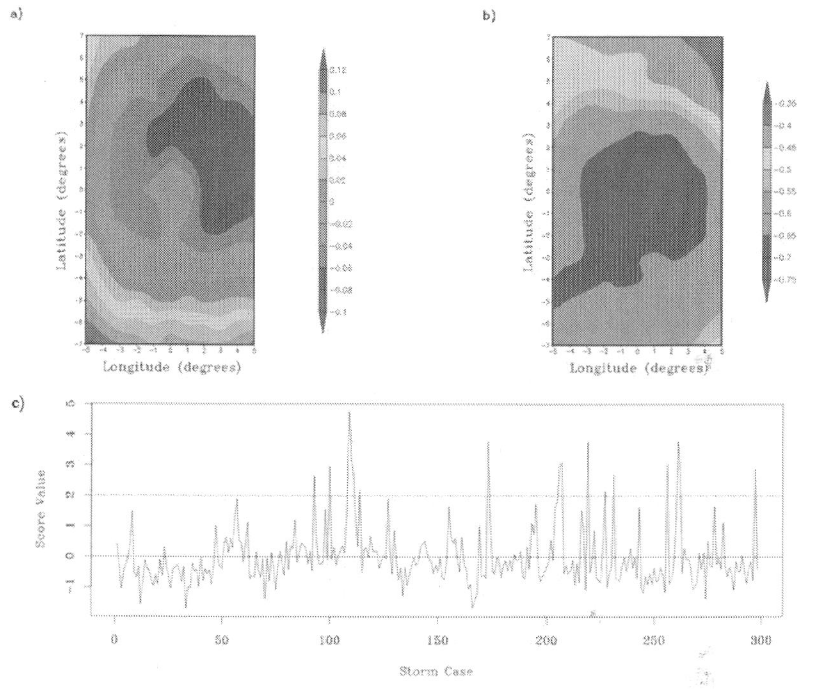

Figure 3. S-mode RPC 1 loading patterns for geopotential height at 850 hPa (panel a) and specific humidity at 850 hPa (panel b) with respect to latitude/longitude relative to the storm center, as well as the associated RPC score time series (panel c). The product of the loading map and its time-series value for a case are in units of standard anomalies.

	S-mode score extremes (±) 2 SDs above mean			
	+2 SD RI	**−2 SD RI**	**+2 SD non-RI**	**−2 SD non-RI**
RPC 1	6	0	10	0
RPC 2	3	4	5	4
RPC 3	1	5	0	8
RPC 4	3	4	4	5
RPC 5	2	7	4	4
RPC 6	1	6	2	2

Table 3. RPC score values exceeding (±) 2 standard deviations above/below the mean.

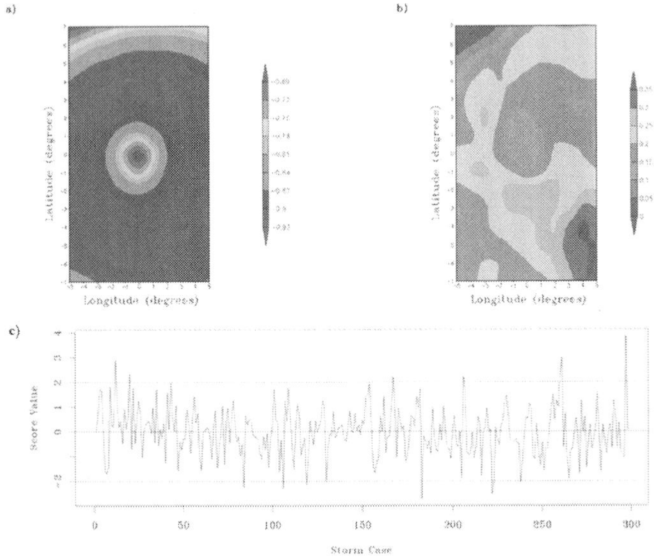

Figure 4. Same as Figure 3, but for RPC 2.

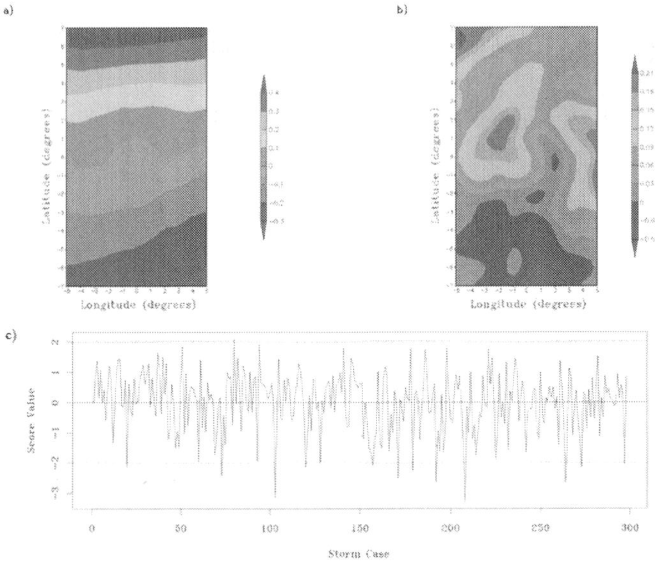

Figure 5. Same as Figure 3, but for RPC 3.

Through this analysis, the inherent difficulty of classifying RI and non-RI storms is apparent, as the base-state fields considered seem to be equally present in RI and non-RI events for all RPCs. Despite this, the results suggest some modest classification ability of RI and non-RI events through the temperature and moisture patterns, as well as variables more indicative of environmental interaction (e.g. vorticity and static stability) as the largest influences on TC RI processes.

3.2. RI and non-RI T-mode composites

T-mode composites of the base-state meteorological fields, as well as derived fields including: divergence, relative vorticity, vertical speed and directional wind shears (see [18] for clarification on the difference), equivalent potential temperature, and static stability (as defined in [19]) were formulated next. The analysis below is broken down by variable.

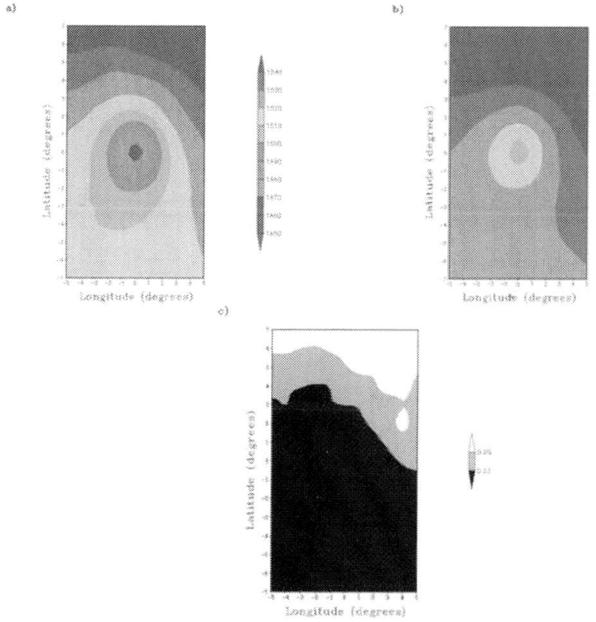

Figure 6. Geopotential height (m), with respect to latitude/longitude relative to the storm center, composites at 850 hPa using 40kt/24-hours definition for cluster 2 RI (panel a) and cluster 3 non-RI (panel b). Permutation tests results (panel c–shaded areas are significant at α=0.05 or less) revealed nearly the entire map as significant in discriminating between RI and non-RI events.

3.2.1. *Geopotential height and mean sea level pressure characteristics*

The map types for RI and non-RI systems revealed a better lower to mid-level structure, with lower heights for a larger radius overall for RI systems. This suggests the RI core is physically distinct from its surrounding environment. In general, for all RI definitions at all height levels, the highest heights are in the northeast quadrant quadrant of the composites, with low-levels for all RI definitions also exhibiting higher heights around the core, indicative of deeper convection (**Figure 6**). In the mid-levels and low-levels for 30kt/24-hours (four out of the six RI clusters) and 40kt/24-hours (two out of the five RI clusters) definitions, all of the RI clusters contain lower heights over the storm core for a larger radius. Map types for MSLP reveal instances when RI composites exhibit a smaller diameter of lower MSLP over the storm center (cluster 6 for RI using 30kt/24-hours and cluster 2 for 40kt/24-hours) with tighter gradients. Comparing these results to non-RI composites, three of the seven clusters maintain a uniform appearance (30kt/24-hours) or even mirror a traditional midlatitude trough/ridge pattern (in one non-RI map type). It is important to note that two non-RI cases using the 40kt/24-hours definition had larger regions of lower MSLP, which is explainable given the frequency of strong (category 3 or 4) non-RI storms associated with this RI definition (12% of the non-RI dataset). Regardless, the dominant pattern among all clusters shows a tighter gradient in low-level geopotential height and MSLP surrounding the TC core in RI systems. Permutation testing revealed a magnitude difference most apparent with MSLP composites for all RI definitions, where RI cases are exhibiting a statistically significantly larger radius of lower heights and pressures than for the non-RI systems, especially for 25kt and 30kt definitions. These results are supported by permutation test results for geopotential heights, which reveal the storm center in the low- and mid- levels as statistically significant at the 95% level in distinguishing between RI and non-RI storms (**Table 4**). It is also notable that the region of significance for MSLP increases as the wind definition increases. In other words, the 40kt/24-hours has the entire permutation map exhibiting statistical significance suggesting geopotential heights are more distinct; however, this could be an artifact of the 40kt definition containing category 4 and 5 storms making up 70% of the dataset versus non-RI containing at most category 4 (4%).

	T-mode analysis for geopotential height and MSLP					
	25kt/24-hours		30kt/24-hours		40kt/24-hours	
Variable	Mag. difference (%)	Permutation significance (%)	Mag. difference (%)	Permutation significance (%)	Mag. difference (%)	Permutation significance (%)
Geo. height 850 hPa	<5	11	<5	49	<5	**83**
Geo. height 700 hPa	<5	5.5	<5	19	<5	42
Geo. height 500 hPa	<5	13	<5	30	<5	12
Geo. height 200 hPa	<5	53	<5	**95**	<5	**96**
MSLP	<5	36	<5	**92**	<5	**100**

Percent significance values greater than 70 are bolded.

Table 4. T-mode analysis results for geopotential height and MSLP. Magnitude difference (%) for RI greater than non-RI composites and percent significance results from permutation tests for each variable examined for each RI definition.

http://dx.doi.org/10.5772/63988

Variable	T-mode analysis for thermodynamic variables					
	25kt/24-hours		30kt/24-hours		40kt/24-hours	
	Mag. difference (%)	Permutation significance (%)	Mag. difference (%)	Permutation significance (%)	Mag. difference (%)	Permutation significance (%)
Spec. humidity 850 hPa	<5	39	<5	**77**	8	**90**
Spec. humidity 700 hPa	6	33	10	69	12	67
Spec. humidity 500 hPa	14	33	13	55	18	38
Spec. humidity 300 hPa	27	40	21	51	22	44
Equiv.pot. temp. 850 hPa	<5	47	<5	**90**	<5	**93**
Equiv. pot. temp. 700 hPa	<5	50	<5	**90**	<5	**94**
Equiv. pot. temp. 500 hPa	<5	56	<5	**90**	<5	**95**
Equiv. pot. temp. 300 hPa	<5	**98**	<5	**100**	<5	**100**
Static stability 850 hPa	<5	2.4	<5	0.6	<5	19
Static stability 700 hPa	<5	16	<5	30	<5	82
Static stability 500 hPa	7.5	**79**	<5	**96**	6	**95**
Latent heat flux	16	58	14	35	11	29
CAPE	<5	46	<5	53	<5	62
CIN	<5	18	<5	5.5	<5	9.1
Skin temp	<5	70	<5	67	<5	60

Percent significance values greater than 70 are bolded.

Table 5. Same as **Table 4**, but for thermodynamic variables.

3.2.2. Thermodynamic characteristics

Specific humidity (25kt/24-hours and 30kt/24-hours) throughout the atmospheric profile contain larger magnitudes (see **Table 5**) for a greater diameter around the storm center and in the northeast quadrant quadrant for RI cases. RI TCs also contain maximum magnitude over the storm center in the mid- and upper- levels, or in the northeast quadrant quadrant, compared to non-RI cases which see a shift of the maximum magnitude towards the ENE region for 25kt/24-hours and 30kt/24-hours definition (**Figure 7**). Cross sections show drier air infiltrating through the inflow regions of the non-RI storm (west side of latitudinal cross section for 25kt/24-hours definition) compared to a more even distribution for RI clusters on either side of the storm center (**Figure 8**).

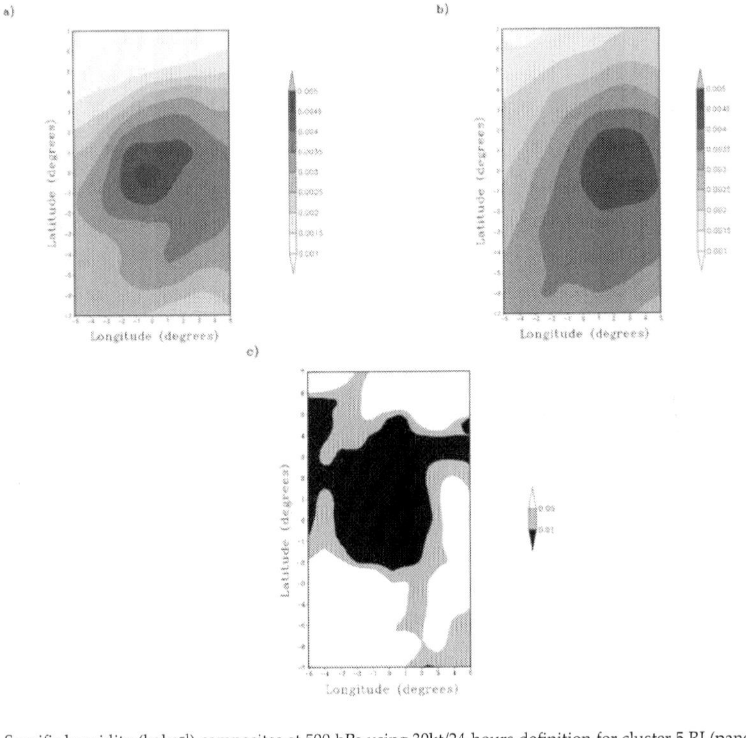

Figure 7. Specific humidity (kgkg^{-1}) composites at 500 hPa using 30kt/24-hours definition for cluster 5 RI (panel a) and cluster 6 non-RI (panel b). Permutation tests results (panel c–shaded areas are significant at α=0.05 or less) revealed the storm center as significant in discriminating between RI and non-RI events.

Equivalent potential temperature (θ_e) fields show similar magnitudes among RI and non-RI cases, although the radius of maximum θ_e is larger with the RI map type, suggesting the potential energy over the storm center is the important feature here. Additionally, the θ_e field is largely symmetric around the storm center for RI map types (**Figure 9a**). However, for the non-RI using the 25kt/24-hours definition, the θ_e field is non-symmetric, instead showing a tilted core in the composite fields (**Figure 9b**). This tilt suggests a cutting off of the moisture source over the storm center, especially in the mid- and upper- levels. These results do not hold up as well for the 30kt/24-hours and 40kt/24-hours RI definitions, as there are non-RI composites which show symmetric latitudinal θ_e cross sections. While the more intense non-RI TCs have clustered together, distinguishing them from the non-RI group itself, it hinders classification ability using these RI definitions. Permutation tests revealed that at all pressure levels, for all RI definitions, the region directly over the storm center, as well as the inflow region for 25kt and 30kt definitions, is statistically significant at the 95% level in discriminating RI versus non-RI systems.

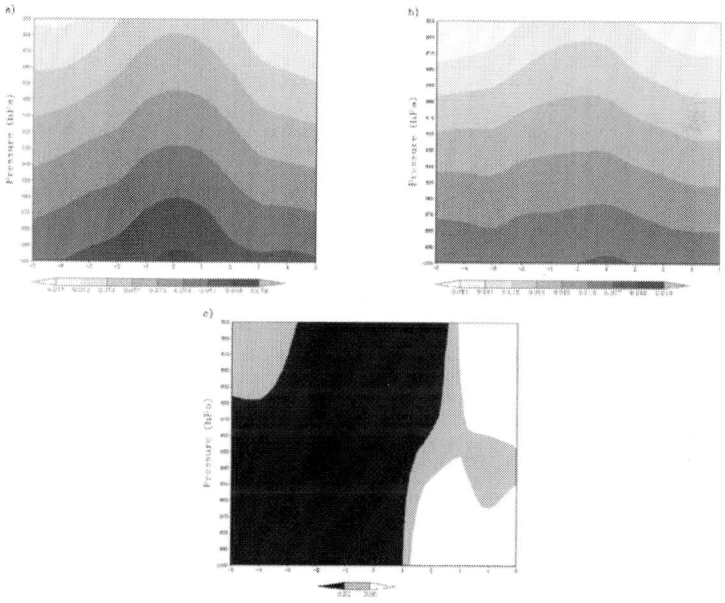

Figure 8. Latitudinal cross section for specific humidity (kgkg^{-1}) composites 1000–850 hPa using the 25kt/24-hours definition for cluster 5 RI (panel a) and cluster 3 non-RI (panel b). Permutation tests (panel c–shaded areas are significant at α=0.05 or less) revealed the storm center and inflow region as statistically significantly different.

Static stability at 500 hPa, on average revealed magnitudes were approximately the same for all definitions between RI and non-RI storms. However, RI clusters, which contained stronger TCs (i.e. category 4 and 5s), had a closed off maximum static stability center over the core of the storm for all RI definitions (**Figure 10**). Permutation tests confirm stability in the mid-levels as statistically significant in discriminating between RI and non-RI systems for nearly the entire storm domain. Notably, for the 40kt/24-hours definition, the storm center in the low-levels was also significant, but is likely a result of higher magnitudes (i.e. category 5 cases) for these RI events.

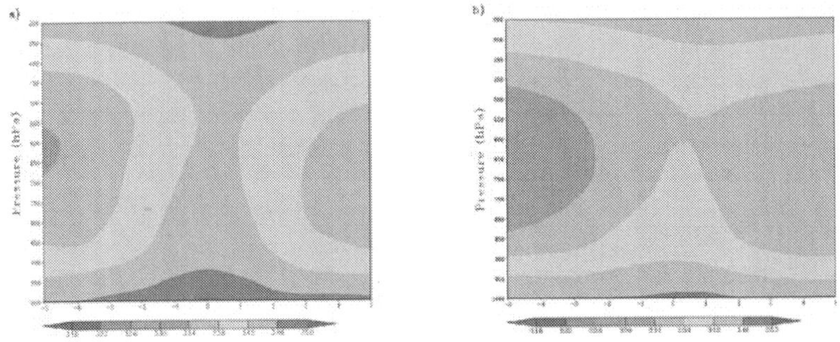

Figure 9. Latitudinal cross section composites for equivalent potential temperature (K) 1000–300 hPa using the 25kt/24-hours definition for cluster 5 RI (panel a) and cluster 3 non-RI (panel b).

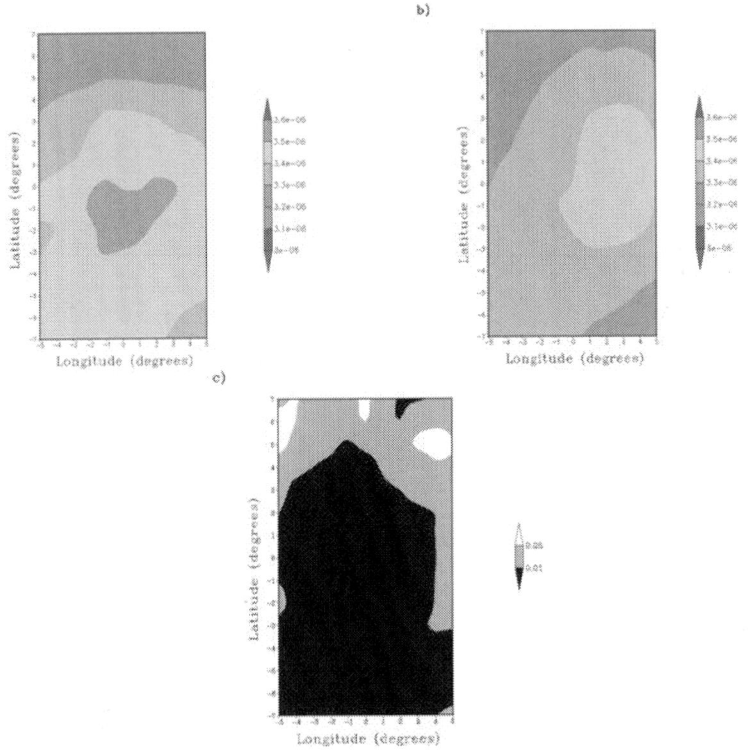

Figure 10. Static stability (K/Pa) composites at 500 hPa using 40kt/24-hours definition for cluster 2 RI (panel a) and cluster 3 non-RI (panel b). Permutation tests (panel c–shaded areas are significant at α=0.05 or less) revealed magnitude discrimination for nearly the entire storm domain.

http://dx.doi.org/10.5772/63988

3.2.3. Kinematic characteristics

The first kinematic field considered was upper-level (200 hPa) divergence. Divergence (25kt/24-hours and 30kt/24-hours) showed RI and non-RI clusters similar in both magnitude and region of greatest divergence in the northeast quadrant quadrant of the systems (**Figure 11**); however, the 40kt/24-hours definition revealed RI systems had 30% larger magnitude near the storm center and in the northeast quadrant quadrant. RI cases, for all definitions, tended to have a larger coverage area of the composite exhibiting divergence, despite the similarities in the spatial orientation of the divergence on the composite maps. Permutation tests supported the conclusion that divergence magnitude (**Table 6**), rather than spatial orientation, was the distinguishing characteristic between RI and non-RI storms at a 95% significance level.

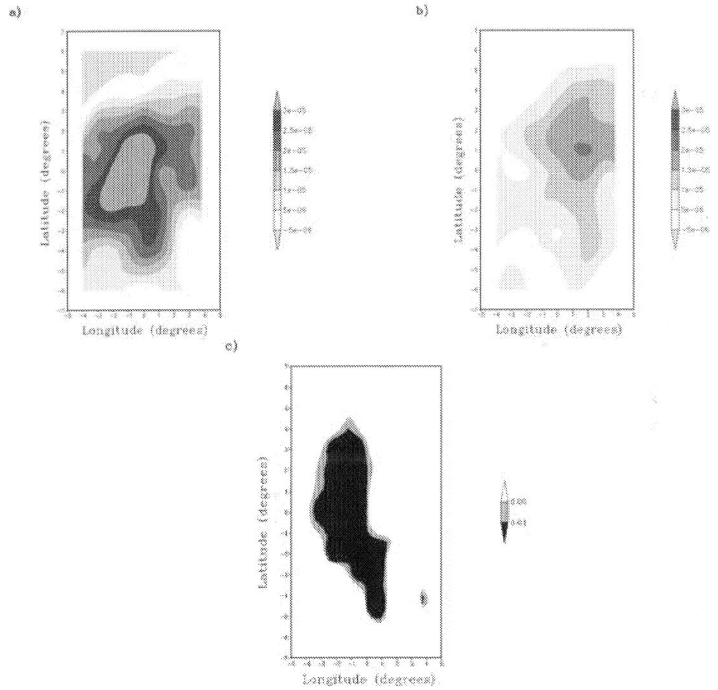

Figure 11. Divergence (s^{-1}) composites at 200 hPa using 30kt/24-hours definition shows an example of how cluster 5 RI (panel a) and cluster 6 non-RI (panel b) clusters are different in both magnitude and region of greatest divergence in the northeast quadrant quadrant of the systems. Permutation tests (panel c–shaded areas are significant at α=0.05 or less) revealed a region over the storm center as significant.

For relative vorticity, using both the 25kt/24-hours and 30kt/24-hours definitions, positive vorticity is noted in three out of the six RI clusters in proximity to storm center in the upper levels, which is notably absent from non-RI cluster map types. For the 40kt/24-hours definition, three out of five RI map types exhibited this feature as well, while only two out of seven non-RI map types showed the same positive vorticity area (**Figure 12**). Vorticity magnitudes were larger for RI TCs with all map types (at all levels) and definitions, and also the vorticity gradient near the center was steeper within the RI system versus the non-RI. The only exception was at 700mb, in which vorticity features are similar in both RI and non-RI. Permutation tests show that over the storm center, for all three pressure levels, all RI definitions, had a 95% level of significance in distinguishing RI from non-RI cases. Notably, the area of statistical significance around the storm center is larger in the mid- levels.

	T-mode analysis for kinematic variables					
	25kt/24-hours		**30kt/24-hours**		**40kt/24-hours**	
Variable	**Mag. difference (%)**	**Permutation significance (%)**	**Mag. difference (%)**	**Permutation significance (%)**	**Mag. difference (%)**	**Permutation significance (%)**
Vorticity 700 hPa	17	12	35	11	28	16
Vorticity 500 hPa	40	19	22	15	38	19
Vorticity 200 hPa	100	22	100	17	28	14
Divergence 200 hPa	<5	23	<5	21	30	21
Directional shear 850–200 hPa	<5	12	<5	3.4	<5	8.5
Speed shear 850–200 hPa	47	**76**	30	**81**	33	**75**

Percent significance values greater than 70 are bolded.

Table 6. Same as **Table 4**, but for kinematic variables.

Map types of vertical speed shear (850–200 hPa – **Figure 13**), thought to be undesirable for RI to occur, revealed weaker 200 hPa winds than the 850 hPa winds within the RI. This is indicative of a closed off environment around the core for the RI systems to a greater degree than the non-RI. Permutation tests revealed all but the southwest quadrant to be statistically significant at the 95% level at discriminating RI from non-RI systems for all RI definitions.

3.2.4. Non-significant variables

CAPE, CIN, vertical velocity at 850 hPa, latent heat flux, sensible heat flux, static stability at 850- and 700-hPa, and skin temperature composites, while all examined, did not reveal

http://dx.doi.org/10.5772/63988

meaningful differences with regards to spatial orientation or magnitude for distinguishing between RI and non-RI cases. While some of the magnitudes were greater for RI clusters containing stronger systems (i.e. category 4 and 5 TCs), other non-RI clusters exhibited similar magnitudes which consisted mainly of tropical storm strength systems. Latent heat flux for example, revealed that for the 25kt/24-hour definition, more latent heat flux was available throughout the inflow region and around the core of the RI cases. However, with the 30kt/24-hour and 40kt/24-hour definitions, the main distinguishing feature seemed to only be higher magnitudes throughout the atmospheric profile. Otherwise, permutation tests surprisingly revealed the NW and northeast quadrant quadrant of the maps for all RI definitions as statistically significant at the 95% level for both CAPE and skin temperature. This is attributed to land influences of some TCs which were in proximity to land when the greatest intensification occurred. Results confirmed a lack of statistical significance in discriminating RI from non-RI with CIN, but confirmed a decent discrimination of magnitude with latent heat flux and sensible heat flux (**Table 5**). However, again, these results are likely being influenced by the proximity to land of some TCs, which would affect 1000 hPa level results for these variables.

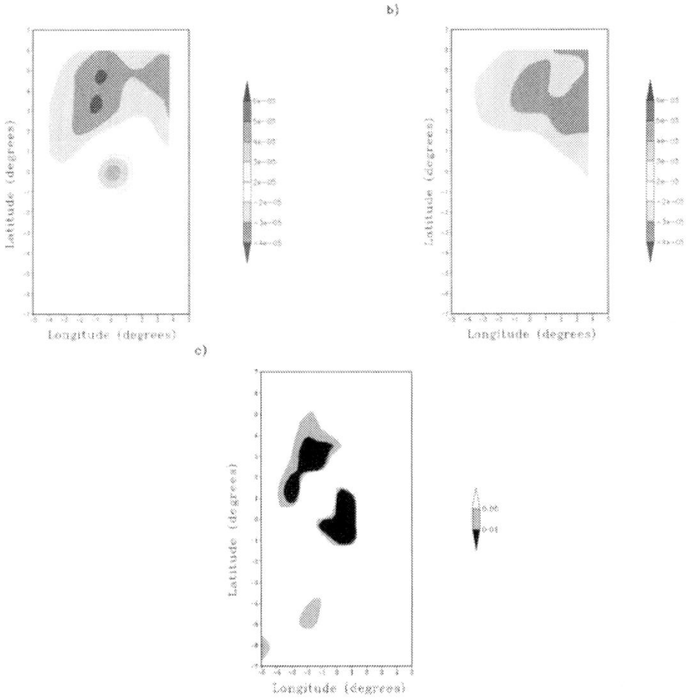

Figure 12. Relative vorticity (s^{-1}) composites at 200 hPa using 40kt/24-hours definition for cluster 2 RI (panel a) and cluster 3 non-RI (panel b). Permutation tests (panel c–shaded areas are significant at α=0.05 or less) revealed the area over the storm center to be statistically significant in distinguishing between the two event types.

Overall, the T-mode analysis revealed the discriminating spatial and magnitude differences between RI and non-RI storms. As suspected through S-mode analysis, moisture and surface temperature patterns, as well as variables indicating environmental influence including geopotential heights in the upper levels, relative vorticity, divergence, and static stability in the mid-levels had the largest influences on TC RI processes.

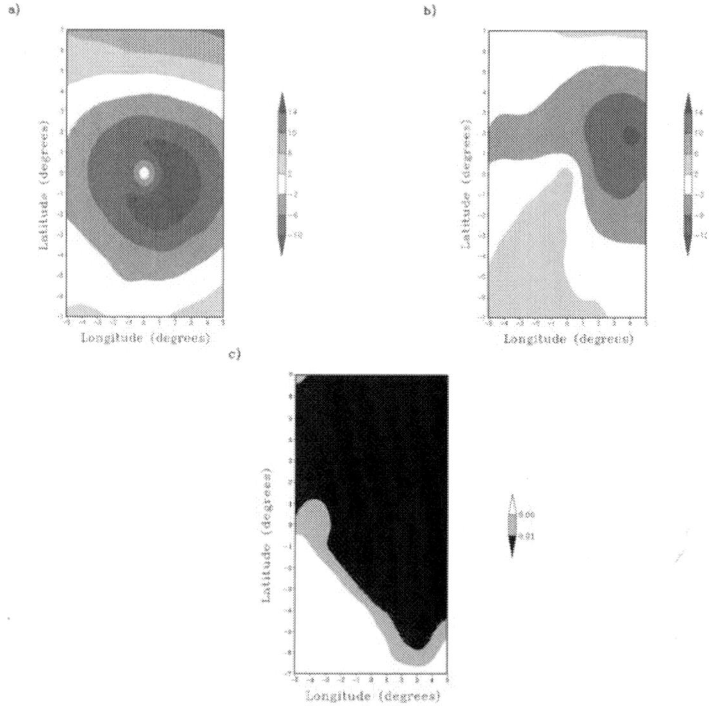

Figure 13. Vertical speed shear (ms^{-1}) composites at 850–200 hPa using 25kt/24-hours definition for cluster 5 RI (panel a) and cluster 3 non-RI (panel b). Permutation tests (panel c–shaded areas are significant at α=0.05 or less) revealed nearly the entire map as statistically significant.

4. Conclusion

Distinguishing meteorological characteristics of RI and non-RI storm structure is critically important in order to improve statistical model prediction of the onset of RI. This research made efforts to continue improvement in identifying relevant large-scale internal dynamics of TCs undergoing RI in the North Atlantic basin, specifically noting important diagnostic variables in three-dimensional space. Base-state, as well as composite derived, meteorological parameters were evaluated through both S-mode and T-mode RPCA for three RI definitions.

Specifically with T-mode, hierarchical cluster analysis techniques were used to formulate map types for RI and non-RI systems. To understand the internal dynamics within these complex systems, variables examined included: geopotential heights, temperature, *u* and *v* wind components, specific humidity, MSLP, CAPE, CIN, latent heat flux, sensible heat flux, surface temperature, vertical velocity at 850 hPa, divergence at 200 hPa, relative vorticity, vertical directional and speed shear, equivalent potential temperature, and static stability.

S-mode analysis results demonstrated the difficulty of establishing characteristic attributes for classifying RI and non-RI storms, as the base-state fields considered were equally present in RI and non-RI events for all RPCs. Two of the six RPC groups contained cases that were indicative of strong, well-structured TCs, exhibiting what you would expect for a sustainable environment for TC continuation and strengthening, regardless of storm type. Whereas, four of the six RPC groups contained cases influenced by baroclinic environmental effects on TCs, which further aids in the positive/negative aspects of environmental influence on TCs. While stronger outflow can lower stability, enhancing the outflow, it can also be detrimental to a TC [20, 21]. Despite this, results indicated a modest classification ability of RI and non-RI events through the temperature and moisture patterns, as well as those variables that would be more indicative of environmental interaction.

T-mode analysis, on the other hand, revealed several important distinguishing spatial features between RI and non-RI systems. Most notably:

- Geopotential heights and MSLP in the low- and mid- levels were statistically significantly different between RI and non-RI systems for all RI definitions, where RI systems maintained lower heights directly over the core and higher heights around the core and in the northeast quadrant quadrant for a larger radius. These results suggest deeper convection for RI systems, re-emphasizing the importance of convective processes around the core of a TC [2, 4, 22, 23].
- Specific humidity throughout the atmospheric profile contained larger magnitudes for a greater diameter around the storm center, as well as in the northeast quadrant quadrant for RI cases (which is in agreement with geopotential heights). Specifically, for the mid- and upper- levels, non-RI cases exhibited a shift of the largest specific humidity magnitudes towards the ENE region and cross sections suggested drier air infiltrating non-RI cases within the inflow regions of the storm for the 25kt/24-hours definition. This was in contrast to a more uniform horizontal and vertical moisture distribution in RI storms.
- Equivalent potential temperature for all RI definitions gave similar results to specific humidity, where the focus of the potential energy was over the storm center for RI cases. Non-RI cases instead exhibited a shift towards the east-northeast (creating the tilted appearance in the composites). This lack of symmetry in non-RI storms for the 25kt/24-hours definition, suggests a cutting off of the moisture and heat source over the storm center, especially in the upper-levels. This region over the storm center was significant in discriminating and indicates enhanced eyewall convection and TC intensity [23].
- Static stability for all RI definitions revealed RI systems were statistically significantly more stable over the storm center in the mid-levels than non-RI systems. This result implies a

resistance to upward vertical motion, forcing subsidence over the storm center [5, 20], but also resistance to adverse effects of the environment, such as vertical shear and Rossby penetration depth, preventing tilting of the TC and allowing for maintenance of the vertical thermal structure [24, 25].

- Divergence at 200 hPa (25kt/24-hours and 30kt/24-hours) showed RI and non-RI cluster composites similar in both spatial location and magnitude of greatest divergence in the northeast quadrant quadrant for the two types of systems. However, RI composites tended to have a statistically significantly larger magnitude.
- Relative vorticity at 200 hPa for the 25kt/24-hours and 30kt/24rs revealed three RI clusters contained an upper-level area of positive vorticity over the storm center. This feature only appeared for the 40kt/24-hours definition for non-RI systems (which is attributed to occurrence of category 4 TC events). Throughout the mid- and upper- levels, RI cases had significantly higher magnitudes over the storm center, indicating a stronger spin.
- Shear, often thought to hinder TC intensification by creating asymmetry in eyewall convection resulting in a loss of the warm core at upper levels through tilting [24–27], revealed vertical speed shear had a much larger area of statistical significance in discriminating RI from non-RI systems for all RI definitions, compared to vertical directional shear currently used in operational forecasts [1].
- CAPE and skin temperature did not reveal any distinguishing feature between RI and non-RI cases through composite analysis. However, permutation tests suggested the NW and northeast quadrant quadrant of the maps for all RI definitions as statistically significant at distinguishing between storm types for both. Latent heat flux, fundamental in the maintenance of convection and increasing kinetic energy [23, 28], showed that RI systems have higher magnitudes for a larger area over the core, for all RI definitions, and throughout the inflow region for the 25kt/24-hours and 30kt/24-hours definitions. However, land masses could be influencing results at the 1000 hPa pressure level for these three, and the other surface variables; therefore, distinguishing whether these fields are different among RIs and non-RIs remains unclear.

While results of the RPCA analysis confirm previous findings such as the importance of moisture supply, stability within the core, and stronger relative vorticity for RI systems, it also argues against research findings suggesting magnitude is the main distinguisher between RI and non-RI events [29]. Results presented suggest the symmetry of the equivalent potential temperature and specific humidity profiles throughout the atmospheric column, as well as the storm-centered placement of these variables, and stability, directly over the inner-core (instead of shifted to the east-northeast as with several non-RI composites given lower RI definitions) are significant in discrimination of these event types. While there were some shortcomings, such as proximity to land potentially influencing results in the low levels and the inability to fully resolve the inner-core due to model resolution, the results provide a framework of diagnosis for RI processes within TCs. This framework, combined with an improved statistical modelling scheme, will ideally be of use for improving TC intensity forecasts in operational meteorology.

http://dx.doi.org/10.5772/63988

Acknowledgements

This work was funded through the Northern Gulf Institute by NOAA Grant #NA11OAR4320199 at Mississippi State University.

Author details

Alexandria Grimes and Andrew E. Mercer*

*Address all correspondence to: a.mercer@msstate.edu

Northern Gulf Institute, Mississippi State University, Mississippi State, MS, USA

References

[1] Kaplan J, Rozoff CM, DeMaria M, Sampson CR, Kossin JP, Velden CS, Cione JJ, Dunion JP, Knaff JA, Zhang JA, Dostalek JF. Evaluating environmental impacts on tropical cyclone rapid intensification predictability utilizing statistical models. Weather and Forecasting. 2015 Oct;30(5):1374-96.

[2] Kaplan J, DeMaria M. Large-scale characteristics of rapidly intensifying tropical cyclones in the North Atlantic basin. Weather and Forecasting. 2003 Dec;18(6):1093-108.

[3] Hendricks EA, Peng MS, Fu B, Li T. Quantifying environmental control on tropical cyclone intensity change. Monthly Weather Review. 2010 Aug;138(8):3243-71.

[4] Rozoff CM, Velden CS, Kaplan J, Kossin JP, Wimmers AJ. Improvements in the probabilistic prediction of tropical cyclone rapid intensification with passive microwave observations. Weather and Forecasting. 2015 Aug;30(4):1016-38.

[5] Vigh JL, Schubert WH. Rapid development of the tropical cyclone warm core. Journal of the Atmospheric Sciences. 2009 Nov;66(11):3335-50.

[6] Sitkowski M, Barnes GM. Low-level thermodynamic, kinematic, and reflectivity fields of Hurricane Guillermo (1997) during rapid intensification. Monthly Weather Review. 2009 Feb;137(2):645-63.

[7] Rozoff CM, Kossin JP. New probabilistic forecast models for the prediction of tropical cyclone rapid intensification. Weather and Forecasting. 2011 Oct;26(5):677-89.

[8] Kaplan J, DeMaria M, Knaff JA. A revised tropical cyclone rapid intensification index for the Atlantic and eastern North Pacific basins. Weather and forecasting. 2010 Feb; 25(1):220-41.

[9] Jarvinen BR, Neuman CJ, Davis MA. A tropical cyclone data tape for the North Atlantic basin. NOAA Tech. Memo. NWS NHC-22. 1988. Mar: 24 pp.

[10] Hamill TM, Bates GT, Whitaker JS, Murray DR, Fiorino M, Galarneau Jr TJ, Zhu Y, Lapenta W. NOAA's second-generation global medium-range ensemble reforecast dataset. Bulletin of the American Meteorological Society. 2013 Oct;94(10):1553-65.

[11] Mercer AE, Shafer CM, Doswell III CA, Leslie LM, Richman MB. Synoptic composites of tornadic and nontornadic outbreaks. Monthly Weather Review. 2012 Aug;140(8): 2590-608.

[12] Grimes A, Mercer AE. Synoptic-scale precursors to tropical cyclone rapid intensification in the Atlantic Basin. Advances in Meteorology. 2015 May 27; 2015: 16pp.

[13] Michael B R, Andrew E M, Lance M L, Charles A III D, Chad M S. High dimensional dataset compression using principal components. Open Journal of Statistics. 2013 Oct 9;2013.

[14] Richman MB. Rotation of principal components. Journal of climatology. 1986 Jan 1;6(3): 293-335.

[15] Richman MB, Lamb PJ. Climatic pattern analysis of three-and seven-day summer rainfall in the central United States: some methodological considerations and a regionalization. Journal of Climate and Applied Meteorology. 1985 Dec;24(12):1325-43.

[16] Wilks DS. Statistical Methods in the Atmospheric Sciences. Oxford: Academic press; 2011.

[17] Rousseeuw PJ. Silhouettes: a graphical aid to the interpretation and validation of cluster analysis. Journal of computational and applied mathematics. 1987 Nov 1;20:53-65.

[18] Markowski P, Richardson Y. On the classification of vertical wind shear as directional shear versus speed shear. Weather and Forecasting. 2006 Apr;21(2):242-7.

[19] Bluestein, H. Synoptic Dynamic Meteorology in Midlatitudes: Principles of Kinematics and Dynamics, Vol. 1. 1st ed. New York, NY: Oxford University Press; 1992. 430 p.

[20] Holland G, Merrill RT. On the dynamics of tropical cyclone structural changes. Quarterly Journal of the Royal Meteorological Society. 1984 Jul 1;110(465):723-45.

[21] Molinari J, Vollaro D. External influences on hurricane intensity. Part I: outflow layer eddy angular momentum fluxes. Journal of the Atmospheric Sciences. 1989 Apr;46(8): 1093-105.

[22] Rogers R. Convective-scale structure and evolution during a high-resolution simulation of tropical cyclone rapid intensification. Journal of the Atmospheric Sciences. 2010 Jan;67(1):44-70.

[23] Li Q, Wang Y, Duan Y. Impacts of evaporation of rainwater on tropical cyclone structure and intensity—a revisit. Journal of the Atmospheric Sciences. 2015 Apr;72(4):1323-45.

[24] DeMaria M. The effect of vertical shear on tropical cyclone intensity change. Journal of the Atmospheric Sciences. 1996 Jul;53(14):2076-88.

[25] Jones SC. The evolution of vortices in vertical shear. I: initially barotropic vortices. Quarterly Journal of the Royal Meteorological Society. 1995 Apr 1;121(524):821-51.

[26] Wang Y, Holland GJ. Tropical cyclone motion and evolution in vertical shear. Journal of the Atmospheric Sciences. 1996 Nov;53(22):3313-32.

[27] Frank WM, Ritchie EA. Effects of vertical wind shear on the intensity and structure of numerically simulated hurricanes. Monthly Weather Review. 2001 Sep;129(9):2249-69.

[28] Ma Z, Fei J, Huang X, Cheng X. Contributions of surface sensible heat fluxes to tropical cyclone. Part I: evolution of tropical cyclone intensity and structure. Journal of the Atmospheric Sciences. 2015 Jan;72(1):120-40.

[29] Kowch R, Emanuel K. Are special processes at work in the rapid intensification of tropical cyclones?. Monthly Weather Review. 2015 Mar;143(3):878-82.

Chapter 3

Evaluating the Progress of Atmospheric Research in Understanding the Mechanics Behind Tropical Cyclone-Induced Tornadogenesis

Jordan L. Rabinowitz

Additional information is available at the end of the chapter

http://dx.doi.org/10.5772/64142

Abstract

As tropical cyclones make landfall along coastlines all over the world, havoc is wreaked on families, businesses, economies, etc. A generally overlooked topic is one of the inherent dangers in tropical cyclone landfalls. These can produce tornadoes in the spiral rain bands coming ashore by encountering increasingly higher frictional convergence. The second component will be assessing which other parameters could be analyzed on a synoptic timescale to evaluate how we can potentially improve longer term predictions of tropical cyclone (TC)-induced tornadogenesis. The analysis showed that the combination of high low-level moisture, vertical temperature gradients, and enhanced vertical wind shear is the key factor that links landfalling tropical cyclones to associated tornadogenesis prior to, during, and/or after the time of landfall. An integral part of the process preceding TC-induced tornadogenesis is the enhanced vertical temperature gradient that develops as the storm maintains warm-core characteristics aloft but develops cold-core characteristics closer to the surface. The examination of forecaster perspectives showed that over the past few years there is strong evidence of forecaster improvement based upon greater cognizance of forecast variables linked to TC-induced tornadoes.

Keywords: tropical cyclones, landfall, tornadoes, frictional convergence, energy helicity index, low-/mid-level lapse rates

1. Introduction

Amidst the added pressures associated with forecasting landfalling tropical cyclones (TCs), tornadic supercells embedded in the outer spiral rain bands are of particular concern to operational forecasters. As the United States has become more industrialized, a larger percentage of the population has settled in coastal regions. As a consequence, the inherent dangers linked to landfalling TCs have become a more prevalent issue as more people are living in areas which are geographically and economically vulnerable to TCs [1, 2]. TC-induced tornadoes can occur during the evening or overnight hours, complicating visual confirmation due to the lack of daylight and/or being rain-wrapped. One of the goals associated with TC forecasting is the accurate anticipation of the timing and location of tornadogenesis with respect to a TC landfall. The tornadoes that do occur in association with TCs are rated based upon the Enhanced Fujita (EF) damage rating scale. A large portion of TC-induced tornadoes are of EF-2 intensity or weaker (maximum winds up to 117 kts), with a small percentage reaching EF-3 (maximum winds between 118 and 143 kts) and even more rarely EF-4 intensity (maximum winds between 144 and 174 kts) [3, 4]. This is reflected in **Figure 1** that illustrates the TC-tornado intensity breakdown based on the analysis in Tropical Cyclone Tornado Records for the Modernized NWS Era [4].

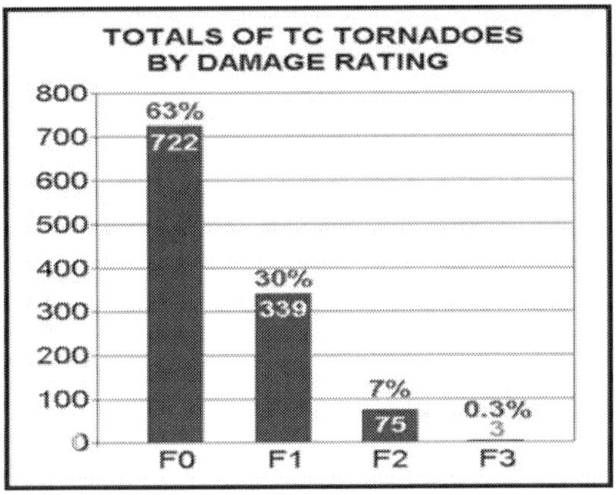

Figure 1. The breakdown of the percentage and number of the 1139 tornadoes that occurred between 1995 and 2009 TC based on their damage rating [4].

The motivation for presenting the content in this chapter is twofold. The first part involves revisiting previous research and assessing how it has benefited the atmospheric research and forecasting communities over the past few decades. The second part is assessing where atmospheric research needs to head moving forward based on the forthcoming results being presented. This chapter is organized as follows. Section 2 consists of an inclusive literature review covering 47 years of TC-induced tornadogenesis research, focusing on relevant statistics and dynamics. Section 3 presents an overview of the data sources that were utilized and the methods by which different contents were evaluated. Section 4 provides a detailed explanation of the results that are divided into two parts: a comprehensive synoptic analysis of 44 United States TC landfall events and detailed analyses of operational forecaster perspectives from the Storm Prediction Center (SPC). The objective of the SPC is to "deliver timely and accurate watch and forecast products/information dealing with severe weather, wildfires, and winter weather for the United States to protect lives and property" (http://www.spc.noaa.gov).

2. Literature review

Past research involving TC-induced tornadogenesis dates back to the middle of the twentieth century. The onset of the satellite era stimulated the atmospheric research community to study mesoscale features embedded within synoptic-scale systems more comprehensively. A paramount issue is the forecasting of tornadoes induced by landfalling TCs along the United States mainland. Over the past six decades considerable progress has been made in understanding the atmospheric dynamics associated with TC-induced tornadogenesis. Early work, which studied TC-induced tornadogenesis, primarily focused on the collection of tornado reports and generating initial theories to explain the variable occurrence of tornadoes during TC landfall events [5, 6].

The work of Smith [6] was composed of compiling a TC-induced tornado climatology using tornado reports from 1955 to 1962. This work proposed a climatological TC tornadogenesis model based on the forward speed of TCs as well as influences from the dynamically driven frictional convergence and strong low-level vertical shear in the northeast quadrant of landfalling TCs. It was noted that some tornado reports are marginal due to the issues such as insufficiently educated spotter reports and/or limited remote sensing capabilities. Nonetheless, this earlier work provided the basis for further progress, starting with Pearson and Sadowski's work [5]. Their work expanded upon [6] by conducting a more detailed assessment of location and timing of tornadogenesis relative to the TC center position. The work of Pearson and Sadowski [5] also provided evidence for tornadogenesis occurring 12 h or more ahead of the arrival of hurricane-force winds. This discovery was monumental as it showed that TC-induced tornadoes occur farther from the center than previously believed.

Novlan and Gray [7] analyzed TC-induced tornadogenesis reports between 1948 and 1972 for the United States and between 1950 and 1971 for typhoons that impacted Japan. The compiled analysis of tornado reports revealed that the number of tornadoes initiated by a given TC was regulated by the rate of cooling aloft, the change in wind speed from the surface to 850 mb, and rising surface pressure values near the TC center upon landfall. It was found that tornadic TCs had weaker winds at the surface and stronger winds aloft, or stronger vertical wind shear [7–9]. In addition, the ambient surface temperatures of tornadic TC environments were lower than those which did not produce tornadoes. The temperatures at 850 mb also remained high in both scenarios, indicating the presence of a low-level cold core that maintained a stronger vertical wind shear profile [7]. This process enhanced cumulus downdraft potential, supporting stronger low-level horizontal wind shear based on a stronger vertical temperature gradient within the tornadic TC environments. Consequently, this induced intense small-scale regions of convergence and rotation, which were objectively associated with the TC-induced tornadogenesis [7].

Moving into the 1980s, Gentry [8] studied TC-induced tornado reports between 1973 and 1980 coupled with considerations based on past research. A core part of his work was appending data from [7] to develop a more comprehensive analysis of tornado reports based on their position relative to the coastline and their distance from TC center positions. A major finding was the majority of tornadoes occurred within 100 km of a given TC center and/or between azimuths of 20° and 120°. The results from [8] reaffirmed that strong vertical wind shear and strong vertical temperature gradients (i.e., cold-core to warm-core changes with height in tornadic TCs) are most responsible for generating environments conducive for tornadogenesis.

In the 1990s, McCaul and Weisman [10] studied composite profiles of temperature, moisture, and wind fields coincident with tornadic TC environments between 1948 and 1986. The premier finding was that helicity (i.e., helical flow) and vertical wind shear parameters were best correlated with TC-induced tornadogenesis. A second major finding was that the number and intensity of TC-induced tornadoes increased in accordance with the increasing TC size and intensity. A final important result from this work was that TCs landfalling along the East Coast produced fewer tornadoes than those that made landfall along the Gulf Coast. Spratt et al. [11] analyzed tornadic mesocyclones associated with two mature TCs that were not close to landfall near the time of tornadogenesis (i.e., Tropical Storm Gordon (1994) and Hurricane Allison (1995)). The primary result was the confirmation of several similarities between TC-induced tornadic cells and Great Plains supercells. In spite of the TC-induced convection being much lower-topped, the ratio of the depth of rotation to storm top was comparable between TC-induced tornadic cells and more common Great Plains supercells. The shallower depth of these rotating cells presented the issue of nondetection based on weaker capabilities of the Weather Surveillance Radar-1988 Doppler (WSR-88D) radar technology. Another important similarity was the persistent nature observed within both Great Plains supercells and TC-induced tornadic supercells.

Heading into the twenty-first century, many breakthroughs advanced the comprehension of dynamics relevant to TC-induced tornadogenesis. The focus of McCaul et al. [12] was studying the remnants of Tropical Storm Beryl (1994) with data from the WSR-88D radar at Columbia, South Carolina, and the National Lightning Detection Network. A major finding was the persistence of offshore low-topped supercells, one of which lasted 11 h and generated multiple tornadoes based on radar data over the course of 6.5 h. In addition, many cases showed a decrease in the frequency of cloud-to-ground (CG) lightning strikes or no lightning activity at all within 30 min of tornadogenesis [12]. Schultz and Cecil [3] conducted a detailed analysis of 1800 TC-induced tornadoes that occurred between 1950 and 2007 to develop an even more extensive TC-tornado climatology as shown in **Figure 2**. Their results supported hypotheses from previous work regarding the differences between inner- and outer-region tornadoes within TC circulations [6, 7]. Another major finding was that the outer-region tornadoes (greater than 200 km from the TC center) exhibited a stronger diurnal signal, with many tornadoes occurring in the afternoon. On the other hand, inner-region tornadoes typically occurred within about 12 h of landfall (without preference for time of day). As stated in [8], the majority of TC tornadoes (60%) occurred within 100 km of the coast. These events include all of the tornadoes in close proximity to the TC core near the time of landfall as well as

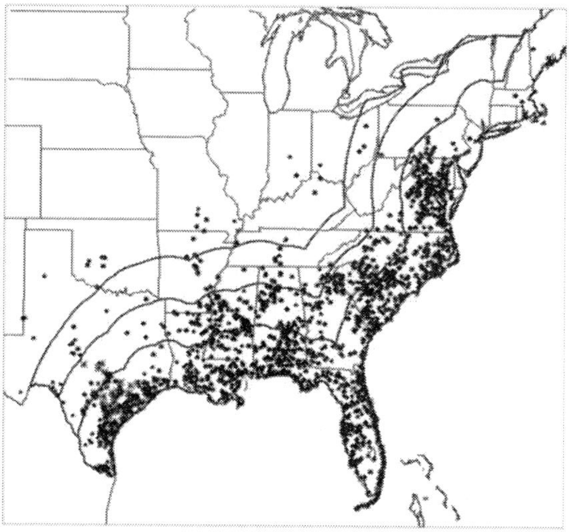

Figure 2. The 1800 tornadoes plotted within the ranges of 200, 400, and 600 km from the respective United States coastlines [3].

tornadoes embedded in rain bands far from the TC center. The last major finding was that the tornadic threat can be found to persist for 2–3 days after landfall and as far as 400–500 km inland from the TC center [3, 13].

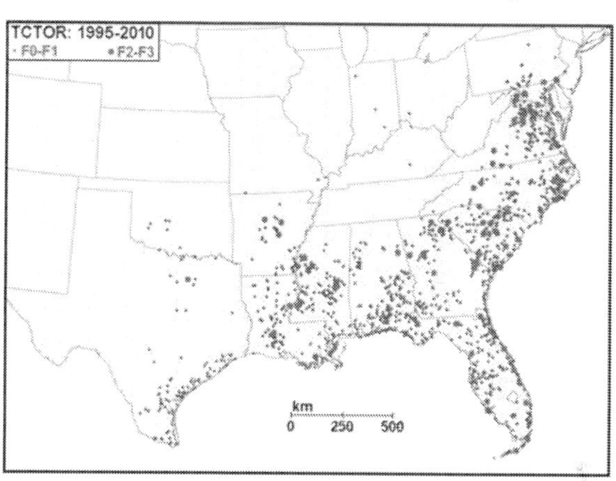

Figure 3. A map of initiation points of 1163 United States TC-induced tornadoes which occurred between 1995 and 2010, damage rating 48 bins as labeled [4].

Edwards [14] studied a TC-tornado dataset dating from 1995 to 2009 from which various graphical and statistical analyses were generated. He conducted a detailed analysis of the position and intensity of 1139 tornadoes that occurred in association with TCs during this 15-year period as shown in **Figure 3**. One revealing result was that the 1139 tornadoes broke down such that there were 722 F0-tornadoes (63.39%), 339 F1-tornadoes (29.76%), 75 F2-torndaoes (6.58%), and 3 F3-tornadoes (0.26%). This statistical breakdown revealed that an overwhelming percentage of the TC-induced tornadoes were characterized by a weaker intensity (i.e., winds of 63 kts or less). The aforementioned TC-induced tornado distribution is comparable to tornado statistics from 1970 to 2002, which were compiled by McCarthy and Schaefer [15]. Within the aforementioned 32-year tornado climatology, the following intensity percentage distribution was found: 39, 36, 19, 5, and 1% for F0, F1, F2, F3, and F4 tornadoes, respectively [15]. The second finding was the apparent peak hours of TC-induced tornado occurrence between 18:00 UTC and 00:00 UTC as reflected in **Figure 4** [14]. This 6-h window in which TC-induced tornadoes occurred most frequently provided evidence for the strong influence of the diurnal cycle on these times of tornadogenesis. This highlighted the importance of diabatic heating and its influence on convective available potential energy (CAPE) values.

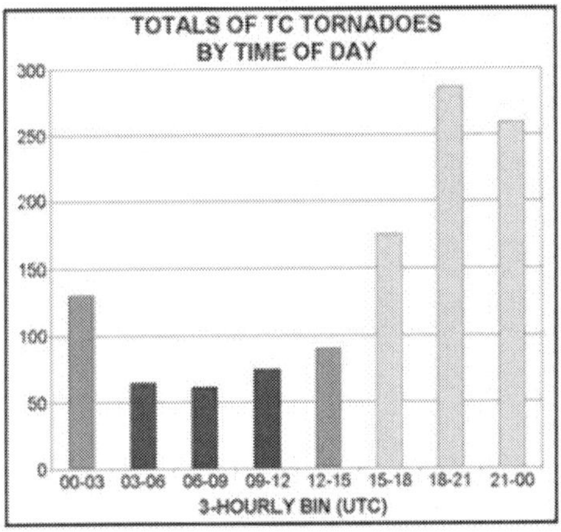

Figure 4. TC-tornado events by UTC time, in 3-hourly groupings. Yellow bars correspond to the local diurnal cycle 19 along the Gulf and Atlantic Coasts, while dark blue bars correspond to the nocturnal cycle. Purple bars correspond to the period of transition between the maximum diurnal cycle impacts and the overnight hours. Periods end in the minute 20 before the labeled times, e.g., "21–00" covers 2100–2359 UTC [4].

Edwards et al. [16] studied a 2003–2011 subset of the Storm Prediction Center (SPC) TC tornado records dataset in conjunction with environmental convective parameters derived from the SPC's hourly mesoscale analysis archive. A key difference observed between TC and non-TC tornado environments was that TC-tornado environments exhibited deeper tropospheric moisture coupled with the reduced lapse rates (near moist adiabatic) and lower CAPE. There was also a proposed objective, which is to study TC-tornado environments more consistently

http://dx.doi.org/10.5772/64142

in order to provide higher-quality data in real time, which may improve the overall effectiveness of operational forecasts in TC-tornado prone locations.

3. Data and methods

A core element of this chapter was analyzing 44 potential TC-tornado events between 2001 and 2015 across the Tropical Atlantic basin. An event was designated as any 24-h period in which a TC was near (within ~160 km) or making landfall along the United States Gulf Coast or East Coast. The critical part of the data collection involved differentiating TC tornadoes vs. tornadoes associated with other mesoscale and/or synoptic-scale systems. This was accomplished by comparing the location and track of a given TC with the positions of reported tornadoes. The final step was analyzing the evolution of available radar data to confirm or reject the validity of reported tornadoes based on the timing of the reports with respect to TC progression.

The first major component of this work was conducting comprehensive synoptic analyses of the lower, middle, and upper levels (as well as an array of model-derived products). These synoptic analyses focused on specified mandatory levels (i.e., 300, 500, 700, 850 mb, and the surface) in conjunction with several model-derived products (see below) from the SPC's hourly mesoscale analysis archive. It is important to note that these model-derived products became fully available from 18 October 2005. Between 3 May 2005 and 17 October 2005, data were only available from 17:00 UTC to 03:00 UTC on the following day (11 out of 24 h per event), limiting the number of consecutive hours available for assessment during that 5-month timeframe.

The model-derived products being analyzed include 0–1-km energy helicity index (EHI), 0–3-km EHI, 100 mb mean parcel lifted condensation level (LCL) height (meters AGL), 300 mb height/divergence/wind, 500 mb height and vorticity, 500, 700, and 850 mb heights/temperatures/winds (measured in decameters, degrees Celsius, and knots), 850 mb moisture transport vector (measured as the product of the wind speed (m/s) and the mixing ratio (g/g) at 850 mb), 850 mb temperature advection (measured as the increasing or decreasing temperature trend) (i.e., positive values indicative of warm-air advection vs. negative values indicative of cold-air advection)), Bulk Richardson Number (BRN) shear (m^2/s^2), effective bulk shear (kts), mean sea-level (MSL) pressure, and precipitable water (inches). Among the 44 events, only 26 events have near/fully complete data sets (June 2005 through June 2015). In addition, some data were missing for 12 events between June 2005 and June 2007, limiting data analysis somewhat during that timeframe.

4. Results

4.1. Environmental analysis

A major facet of this work was conducting in-depth synoptic analyses of 44 individual events that occurred between 2001 and 2015. The first part of this analysis focused on the evolution of the upper-level dynamics by looking at 300 mb heights, divergence, and winds. The main focus was assessing upper-level divergence with respect to the number of tornadoes that occurred during each event. The average maximum value for 300 mb divergence among the 44 events was $4.32 \times 10^{-5}\ s^{-1}$, whereas the majority of tornadic TCs had divergence values between 4 and $6 \times 10^{-5}\ s^{-1}$ (see **Table 1**). This suggests that upper-level divergence had at least some influence on tropospheric mass evacuation near convective cells moving ashore by enabling thunderstorm updrafts to reach higher altitudes. However, previous work has established that low- and mid-level dynamics play a larger role in TC-induced tornadogenesis [5, 7, 8, 17].

In examining the middle levels, the analyzed parameters included 500 mb heights/temperatures/winds and 500 mb height and absolute vorticity. Collectively, the 500 mb height, temperature, and wind values had respective average values of 580.35 dm, −4.33°C, and 35.47 kts. For the most tornadic TCs, the majority of 500 mb height values ranged between 5 dm above and below the average value. The 500 mb temperature values had a larger range of variability, with the most tornadic TCs having values range from −8 to 5°C. For the most tornadic TCs, the 500 mb wind values ranged from 25 to 60 kts. The second aspect evaluated at the 500 mb level was 500 mb height and absolute vorticity (for the 15 data-ready cases) for which the average value was $37.47 \times 10^{-5}\ s^{-1}$. It is important to note that the 500 mb vorticity values ranged from 24×10^{-5} to $58 \times 10^{-5}\ s^{-1}$, indicating that greater vorticity was not correlated with increasingly tornadic TCs. For a deeper look into these mid-level parameters, refer **Table 1** for average value comparisons between all of the events and the most tornadic events.

In stepping down to lower levels, 700 mb heights, temperatures, and winds were the focus. For the most tornadic TCs, the average 700 mb height value was calculated to be 306.16 decameters. Yet, among the various case studies, the 700 mb height values range from 292 to 316 dm. This indicated that the most tornadic TCs occurred with both below- and above-average 700 mb heights, suggesting that the proximity of the TC center to the coastline may have played a key role. As a landfalling TC moves inland, the associated pressure gradient weakens and the lowest minimum central pressure rises. This process is a direct consequence of the transient balance between the rate of mass adjustment toward the center (i.e., via cyclonic inflow toward the TC center) and the rate of tropospheric mass evacuation [3, 7]. The most tornadic TCs, occurring with above-average heights, likely unfolded due to rising mid-level heights coupled with a conditionally unstable convective boundary layer during the post-landfall phase of a particular TC [3, 9, 11]. In examining the 700 mb temperatures, the average value was 9.34°C, and most tornadic TCs were close to this value when measured near the height of each event. By analyzing the 700 mb winds, the average value came out to be 39.09 kts. However, during the most tornadic TCs, the 700 mb wind values ranged from 25 to 90 kts, which is higher than the typical values. For these lower level parameters, refer **Table 1** for comparisons of the average values for all of the events as compared to the most tornadic events. During some landfalling TCs in which 30 or more tornadoes were reported, the mean effective bulk shear value was 40 kts; thus, it is plausible that the larger number of tornadoes was due to larger magnitudes of the low-level shear, coupled with the 850 mb wind data and other low-level data discussed below.

Critical TC-induced tornado parameter averages						
Parameter (number of data-ready cases)	300 mb divergence (10^{-5} s^{-1}) (44 cases)	500 mb height (dam) (44 cases)	500 mb temperature (°C) (44 cases)	500 mb wind speed (kts) (44 cases)	500 mb absolute vorticity (10^{-5} s^{-1}) (15 cases)	
Total event average	4.32	580.35	−4.33	35.47	37.47	
Most tornadic event average (10 + tornadoes)	4.67	578.57	−3.57	41.43	39.14	
Parameter (number of data-ready cases)	700 mb height (dam) (44 cases)	700 mb temperature (°C) (44 cases)	700 mb wind speed (kts) (44 cases)	850 mb height (dam) (44 cases)	850 mb temperature (°C) (44 cases)	850 mb wind speed (kts) (44 cases)
Total event average	306.16	9.34	39.1	142.41	17.14	40.91
Most tornadic event average (10 + tornadoes)	303.95	9.57	44.76	139.81	16.9	46.48
Parameter (number of data-ready cases)	0–1 km EHI (26 cases)	0–3 km EHI (26 cases)	100 mb mean parcel LCL height (m AGL) (26 cases)	BRN shear (kts) (26 cases)	Effective bulk shear (kts) (26 cases)	Mean sea-level pressure (mb) (44 cases)
Total event average	2.83	3.31	640.38	106.4	41.74	985.55
Most tornadic event average (10 + tornadoes)	2.75	3.29	625	130	44.17	982

Table 1. Total events average and most tornadic event average values for 300 mb divergence, 500 mb heights/temperatures/winds and absolute vorticity, 700 mb heights/temperatures/winds, 850 mb heights/temperatures/winds, 0–1 and 0–3 km EHI, 100 mb mean parcel LCL height, BRN shear, effective bulk shear, and mean sea-level pressure.

Further, low-level analysis involved examining the 850 mb heights/temperatures/winds, 850 mb moisture transport vector, and 850 mb temperature advection (the last two parameters only having 15 data-ready cases). The average 850 mb height, temperature, and wind values were calculated to be 142.41 dm, 17.14°C, and 40.91 kts. A critical finding was that the 850 mb heights associated with the more tornadic TCs were near or below the average value, indicating that lower 850 mb heights favored more tornadoes, which agreed with the previous work [7]. For the most tornadic TCs, 850 mb temperatures stayed near or below average, illustrating that TCs which maintained lower mid-level temperatures tended to produce more tornadoes. This is a result of those TCs maintaining warmer near-surface temperatures and cooler mid-level temperatures, strengthening vertical wind shear, which is a critical aspect of TC-induced tornadogenesis. This concurs with the previous work finding that TCs that maintain a warm-core structure aloft, while developing a cold-core structure near the surface, develop stronger vertical wind shear [7]. Consequently, this stronger vertical wind shear fosters an environment more conducive for TC-induced tornadogenesis. Finally, 850 mb wind speeds for the larger tornado producing TCs were at or above the average value, indicating that stronger low-level winds are more favorable for tornadogenesis assuming the presence of weaker winds at the surface. For the 850 mb parameters, refer **Table 1** for comparisons of the average values for all of the events as compared to the most tornadic events.

Another important component was the 850 mb moisture transport (flux) vector, which is the product of the wind speed (m/s) and the mixing ratio (g/g) at 850 mb. This product (studied for 15 data-ready cases) showed that within 12 h of TC-tornadogenesis, there was a large quantity of warm/moist air advected into the tornadic regions. This provided abundant water vapor, which has been shown to be crucial for TC tornadogenesis as well as more common Great Plains tornadic supercells [9, 18]. The last 850 mb product being analyzed was 850 mb temperature advection that among the 15 data-ready events had an average value of -2.31×10^{-5}°C s^{-1}. It was found that 850 mb temperature advection among the most tornadic TCs ranged from 10×10^{-5} to -20×10^{-5}°Cs^{-1} near the time of maximum tornadogenesis. More specifically, this meant that during the most tornadic TCs there was either weak warm-air advection or weak/moderate cold-air advection occurring on a 3 h timescale. Taking past research into consideration, it makes sense that the most tornadic TCs occurred with below-average 850 mb temperatures [7]. For cases with lower 850 mb temperatures, more research is needed to better understand why comparable numbers of tornadoes occur when the vertical temperature gradient is weaker.

Other critical components of TC-induced tornadogenesis include the presence of vertical wind shear and more near-surface variables. In regards to the near-surface domain, analyses were conducted for 0–1 km EHI, 0–3 km EHI, 100 mb mean parcel LCL height (measured in meters AGL), BRN shear, effective bulk shear, MSL pressure, and precipitable water depth (all of which had only 26 data-ready cases except for MSL for which data were available for all 44 events). Taking the EHI into consideration, it is worth noting that the representative equation is

$$\mathrm{EHI} = \left(\frac{\mathrm{CAPE}}{\mathrm{storm - relativehelicity}}\right) \times 160{,}000;$$

which is the ratio between CAPE and low/mid-level wind shear represented by storm-relative helicity, which indicates the relative importance of each variable. The average values for 0–1 km EHI and 0–3 km EHI were 2.83 and 3.31, respectively, with the most tornadic TCs having greater than average values for both variables. This indicates the greater value of buoyancy relative to wind shear for tornado producing events. In addition, EHI values larger than one indicate a greater potential for tornadoes up to EF-3 strength, this is a very propitious result that needs to be further evaluated as also discussed by Eastin and Link [19]. The next parameter evaluated was the 100 mb mean parcel LCL height that was given in meters AGL. The average value for the 100 mb mean parcel LCL heights was calculated to be 640.38 m AGL for the 26 data-ready cases. Most tornadic TCs had 100 mb mean parcel LCL heights ranging from 140 m AGL below to 100 m AGL above average. This provided supporting evidence that lower LCL heights are more favorable for TC-induced tornadogenesis (i.e., 100 mb mean parcel LCL heights under 1000 m AGL) [9, 19]. For these near-surface parameters, refer to **Table 1** for comparisons of the average values for all of the events as compared to the most tornadic events.

The last part of the synoptic analysis considered the following variables: BRN shear, effective bulk shear, MSL pressure, and precipitable water depth. BRN shear is the square of the bulk vector difference between the 0 and 500 m AGL mean wind (both pressure weighted) and then multiplied by one half (http://www.spc.noaa.gov). It has been found that higher BRN values are linked to an increasing risk of supercells based on BRN values being defined by the ratio of buoyancy and shear. This supports the findings from the EHI analyses, which is contingent upon the ratio of convective available potential energy (CAPE) and wind shear in the form of storm-relative helicity [11]. The average BRN shear value was 106.4 m^2 s^2, whereas the most tornadic TCs had values ranging between 80 and 220 m^2 s^2. This is important because BRN shear values at or above 40 m^2 s^2 support environments more conducive for TC-induced tornadogenesis [20]. Looking at effective bulk shear, the average value was calculated to be 41.74 kts, whereas the most tornadic TCs had values ranging from 40 to 60 kts. This is important based on previous work finding that supercell development becomes more likely as effective bulk shear increases to 25–40 kts or greater (http:////spc.noaa.gov). Upon inspecting the evolution of precipitable water, it became clear that values in the vicinity of a landfalling TC were consistently over 2.00 in. This concurs with previous work finding that high water vapor content is essential for TC-induced tornadogenesis due to the necessity for deep moisture within the convective boundary layer [8, 9, 16, 19]. For these additional near-surface parameters, refer to **Table 1** for aforementioned average value comparisons.

The final aspect being assessed is mean sea-level pressure, whose average value among the 44 events was calculated to be 985.55 mb. It is important to mention that the lowest mean sea-level pressure values were all calculated when the storm was approaching landfall (i.e., the TC center was within 160 km of the coastline). For the most tornadic TCs, the mean sea-level pressure values ranged anywhere from 949 to 1008 mb. This indicated significant variability among the surface pressures of landfalling TCs, suggesting that a more intense TC at the time of landfall does not directly correlate with the production of more tornadoes. However, previous work has shown that the majority of tornadoes occur with rapidly weakening TCs upon landfall (i.e., TC centers rapidly filling in terms of the net mass adjustment) [7]. For a more detailed comparison of the average values for all of the events as compared to the more tornadic events, refer to **Table 1**.

4.2. Forecast analysis

By examining this issue with a social impact perspective, it is clear that the public's information outlet is contingent upon the communication of operational meteorologists. This section is a detailed analysis of 124 archived mesoscale discussions from the SPC beginning in August 2004. The forecasters responsible for these mesoscale discussions include the following: C. Broyles, G. Carbin, Dr. A. Cohen, S. Corfidi, M. Darrow, S. Dial, R. Edwards, S. Goss, J. Grams, J. Hart, R. Jewell, B. Kerr, E. Leitman, Dr. P. Marsh, M. Mosier, J. Peters, J. Picca, J. Rogers, R.

Thompson, B. Smith, and the late Jonathan Racy. The impetus for studying these archived mesoscale discussions was to look for a trend in forecaster improvement over the last 10–15 years. The primary focus was assessing the timeliness and accuracy of warnings conveyed by forecasters. Mesoscale discussions for events during and after 2004 were chosen since that was when the SPC began publishing graphics to visually illustrate their dialogue. The addition of graphics provided further insight into the thoughts of forecasters during a particular situation. The principal factors being studied were the number of times the following words appeared at least once in a given discussion: high temperature(s) or surface heating, instability (i.e., CAPE (the amount of energy available for convection), SBCAPE(surface-based CAPE → the value of CAPE relative to an air parcel rising from the lower planetary boundary layer), MLCAPE(mixed-layer CAPE → CAPE calculated with values of temperature and moisture from the lowest 100 mb above ground level), DCAPE(downdraft-CAPE → CAPE calculation which estimates the strength of rain and evaporatively cooled downdrafts), wind shear (i.e., bulk shear, low-level shear, sheared profile, etc.), storm relative helicity (SRH) in the 0–1 or 0–3 km layer, and precipitable water (http//www.spc.noaa.gov). The intention was to assess which factors were most critical for TC-induced tornadogenesis from the standpoint of SPC mesoscale forecasters.

Of the 124 mesoscale discussions, 49 discussions (39.52%) had at least one mention of surface heating and/or high temperatures. Then, 47 of the 124 discussions (37.90%) contained at least one mention of instability terminology. Also, 100 of the 124 discussions (80.65%) contained at least one mention of wind shear terminology. In addition, 66 of the 124 discussions (53.23%) contained at least one mention of storm relative helicity terminology. Finally, 7 of the 124 discussions (5.65%) contained at least one mention of precipitable water. Collectively, wind shear was a prominent topic of many discussions, coinciding with the past research findings that strong low-level wind shear is essential for TC-induced tornadogenesis [10, 16].

In addition, the frequent presence of surface heating and/or high temperature wordings, instability terminology, and storm-relative helicity terminology indicated the issues' prevalence to forecasters involved in watch and/or warning coordination [14]. The main point ascertained from the relative frequency of the discussion parameters from 2004 to 2015 is that SPC forecasters clearly have improved during the latter dual-polarization era that began in late mid to late 2012 across much the contiguous United States (particularly along the Gulf and East Coast regions). This margin of improvement is distinguished by a greater presence of wind shear terminology that as previously stated is crucial to TC-induced tornadogenesis. This may suggest that advanced remote sensing capabilities, as in the implementation of dual-polarization radar technology, in concert with high-resolution rapid-scan satellite imagery, have bolstered forecaster analysis quality.

5. Conclusions

Given the growth of the global population, especially those living near coastlines, and the consistent threats associated with landfalling TCs, there is a growing need for improvements to the modeling and forecasting of TC intensities and trajectories. As a nearly 50-year period of research has shown, tremendous progress has been made in understanding the dynamics that govern TC-induced tornadogenesis. Yet, considerable work still awaits the atmospheric science community.

The results of the environmental analysis showed that upper- and mid-level factors help facilitate TC-induced tornadogenesis. In some cases, upper-level divergence appeared to support tropospheric mass evacuation, bolstering upward vertical velocities within convective cells. It is imperative to note that upper-level divergence was not essential based on several cases with 10+ tornadoes occurring without significant upper-level divergence. Inspecting the middle levels, 500 mb heights/temperatures/winds experienced notable variability that was likely due to differences prior to and after landfall (i.e., pre-landfall TCs were associated with lower 500 mb heights, lower 500 mb temperatures, and stronger 500 mb wind speeds, whereas post-landfall tornadic TCs were the opposite coupled with a greater tendency for synoptic interactions). In regards to 500 mb absolute vorticity, larger values were not synonymous with greater tornado production.

Analysis of 700 and 850 mb data revealed that most tornadic TCs occurred with at or below-average heights, near-average temperatures, and above-average winds. The tendency for at or below-average heights coupled by near-average temperatures (i.e., average values being 9.34°C at 700 mb and 17.14°C at 850 mb) backs the presence of a notable vertical temperature gradient that is pivotal in facilitating TC-induced tornadogenesis. Approaching the surface, the variables, i.e., 100 mb mean parcel LCL heights, BRN shear, effective bulk shear, and mean sea-level pressure, showed convincing output (as shown in **Table 1**). Collectively, it is clear that the essential components most responsible for TC-induced tornadogenesis reside within the planetary boundary layer coupled with some relevant mid-level factors.

However, as with many aspects of meteorology, no individual numerical weather prediction product (i.e., a deterministic model forecast) can be the sole product with which we determine the likelihood of a particular TC producing or not producing tornadoes. Rather, it is with the integration of ensemble forecasts and appropriate temporal and/or spatial parameterizations that atmospheric research will fortify additional headway in generating more accurate TC forecasts. By the collaborative efforts of both atmospheric researchers from the United States and other countries around the world, this is the method by which atmospheric science will most productively push forward in this pursuit of knowledge.

Author details

Jordan L. Rabinowitz

Address all correspondence to: jlrxw6@mail.missouri.edu

Department of Soil, Environmental, and Atmospheric Sciences, University of Missouri, Columbia, MO, United States

References

[1] Pielke, R A, Landsea C W. La Niña, El Niño, and Atlantic Hurricane Damages in the United States. Bull. Am. Meteorol. Soc. 1999; 80: 2029–2032. DOI: 10.1175/1520-0477(1999)080<2027%3ALNAENO>2.0.CO%3B2

[2] Pielke, R A, Gratz J, Landsea C W, Collins D, Saunders M A, Musulin R. Normalized Hurricane Damage in the United States: 1900-2005. Nat. Hazards Rev. 2008; 31: 29–38. DOI: 10.1061/(ASCE)1527-6988(2008)9:1(29)

[3] Schultz, L A, Cecil D J. Tropical Cyclone Tornadoes, 1950–2007. Mon. Wea. Rev. 2009; 137: 3471–3483. DOI: 10.1175/2009MWR2896.1

[4] Edwards, R. Tropical cyclone tornadoes: A review of knowledge in research and prediction. Electronic J. Severe Storms Meteor. 2012; 7(6): 1–28.

[5] Pearson, A D, Sadowski A F. Hurricane-induced tornadoes and their distribution. Mon. Wea. Rev. 1965; 93: 461–464. DOI: 10.1175/1520-0493(1965)093<0461%3AHITATD>2.3.CO%3B2

[6] Smith, J S. The Hurricane-Tornado. Mon. Wea. Rev. 1965; 93: 453–457. DOI: 10.1175/1520-0493(1965)093<0453:THT>2.3.CO;2

[7] Novlan, D J, Gray W M. Hurricane-Spawned Tornadoes. Mon. Wea. Rev. 1974; 102: 476–488. DOI: 10.1175/1520-0493%281974%29102<0476%3AHST>2.0.CO%3B2

[8] Gentry, R C. Genesis of Tornadoes Associated with Hurricanes. Mon. Wea. Rev. 1983; 111: 1794–1797:1802. DOI: 10.1175/1520-0493%281983%29111<1793%3AGOTAWH>2.0.CO%3B2

[9] Baker, A K, Parker M D. Environmental Ingredients for Supercells and Tornadoes within Hurricane Ivan. Wea. Forecasting. 2008; 24: 225–226:242. DOI: 10.1175/2008WAF2222146.1

[10] McCaul Jr., E W, Weisman M L. Simulations of Shallow Supercell Storms in Landfalling Hurricane Environments. Mon. Wea. Rev. 1995; 124: 426–427. DOI: 10.1175/1520-0493%281996%29124<0408%3ASOSSSI>2.0.CO%3B2

[11] Spratt, S M, Sharp D W, Welsh P, Sandrik A, Alsheimer F, Paxton C. A WSR-88D Assessment of Tropical Cyclone Outer Rainband Tornadoes. Wea. Forecasting. 1997; 12: 498–499. DOI: 10.1175/1520-0434%281997%29012<0479%3AAAWAOTC >2.0.CO%3B2

[12] McCaul Jr., E W, Buechler D E, Goodman S J, Cammarata M. Doppler Radar and Lightning Network Observations of a Severe Outbreak of Tropical Cyclone Tornadoes. Mon. Wea. Rev. 2003; 132: 1758–1761. DOI: 10.1175/1520-0493%282004%29132<1747%3A DRALNO>2.0.CO%3B2

[13] Edwards, R, Smith B T, Thompson R L, Dean A R. Analyses of Radar Rotational Velocities and Environmental Parameters for Tornadic Supercells in Tropical Cyclones. In: Proceedings of the 37th Conference of Radar Meteorology; 2006. p. 2:5-10.

[14] Edwards, R. Tropical cyclone tornado records for the modernized National Weather Service era. In: Proceedings of the 25th Conference on Severe Local Storms; Denver, CO; 2010.

[15] McCarthy, D, Schaefer J. Tornado trends over the past thirty years. In: Proceedings of the 14th Conference on Applied Meteorology, Seattle, WA; 2004.

[16] Edwards, R, Dean A R, Thompson R L, Smith B T. Convective Modes for Significant Severe Thunderstorms in the Contiguous United States. Part III: Tropical Cyclone Tornadoes. Wea. Forecasting. 2012; 27: 1509–1510:1513-1518. DOI: 10.1175/WAF-D-11-00117.1

[17] McCaul Jr., E W. Buoyancy and Shear Characteristics of Hurricane-Tornado Environments. Mon. Wea. Rev. 1991; 119: 1960:1965:1974-1977. DOI: 10.1175/1520-0493%281991%29119<1954%3ABASCOH>2.0.CO%3B2

[18] Rasmussen, E N, Blanchard D O. A Baseline Climatology of Sounding-Derived Supercell and Tornado Forecast Parameters. Wea. Forecasting. 1998; 13: 1154:1158. DOI: 10.1175/1520-0434(1998)013<1148:ABCOSD>2.0.CO;2

[19] Eastin, M D, Link M C. Miniature Supercells in an Offshore Outer Rain Band of Hurricane Ivan (2004). Mon. Wea. Rev. 2009; 137: 2102. DOI: 10.1175/2009 MWR2753.1

[20] Thompson, R L, Mead C M, Edwards R. Effective Bulk Shear in Supercell Thunderstorm Environments. In: Proceedings of the 22nd Conference on Severe Local Storm; 2014.

Chapter 4

Upper Ocean Physical and Biological Response to Typhoon Cimaron (2006) in the South China Sea

Yujuan Sun, Jiayi Pan and William Perrie

Additional information is available at the end of the chapter

http://dx.doi.org/10.5772/64099

Abstract

The physical dynamic and biological response processes to Typhoon Cimaron (2006) in the South China Sea are investigated through the three-dimensional Regional Ocean Modeling System (ROMS). For sea surface temperatures, ROMS achieves a correlation of more than 0.84, with respect to satellite observations, indicating a generally high level of skill for simulating the sea surface temperature variations during Typhoon Cimaron (2006). However, detailed analysis shows that ROMS underestimates the sea surface temperature cooling and mixed layer deepening because of insufficient mixing in the model simulations. We show that the simulation accuracy can be enhanced by adding a wave-induced mixing term (B_V) to the nonlocal K-profile parameterization (KPP) scheme. Simulation accuracy is needed to investigate nutrients, which are deeply entrained to the oligotrophic sea surface layer by upwelling induced by Typhoon Cimaron, and which plays a remarkable role in the subsequent phytoplankton bloom. Simulations show that the phytoplankton bloom was triggered 5 days after the passage of the storm. The surface ocean was restored to its equilibrium ocean state by about 10–20 days after the typhoon's passage. However, on this time-scale, the resulting concentrations of nitrate and chlorophyll *a* remained higher than those in the pre-typhoon equilibrium.

Keywords: Typhoon Cimaron, SST cooling, mixed layer deepening, phytoplankton bloom, wave-induced mixing

1. Introduction

Tropical cyclones are extremely high wind events generated over tropical oceans, capable of producing strong mixing and entrainment, transient upwelling, and internal waves. The response of upper ocean water to typhoons can be conventionally divided into two stages, the forcing stage and the relaxation stage in [1, 2]. In the first stage, potential energy is injected into the surface ocean by strong typhoon winds. Two significant physical phenomena caused by typhoons are the mixed layer deepening and sea surface temperature cooling in the wake of the storm. Warm water in the ocean surface layers is transported outward from the typhoon center and downward to depths ranging from tens of meters to beyond a hundred meters; cold water upwells from the deeper ocean along the typhoon's passage [3]. The current velocity in the surface mixed layer can reach 2 m s^{-1} or more, responding to the intensity of the tropical storm winds [2]. The relaxation stage following a typhoon's passage is primarily due to the inertial gravity oscillations excited by the storm in its wake, where the ocean adjusts towards a new geostrophic equilibrium state [1, 4–6]. The work of [7] concluded that the inertial oscillations are predominantly locally generated and the surface winds account for a large part of the energy and variability of such oscillations near the ocean surface. The theory of geostrophic adjustment was further reviewed in [8]. The latter theory, which describes the nature of the difference between the baroclinic and barotropic responses of the ocean to a moving storm as caused by the difference in the gravity wave speed, was first pointed out in [9]. The storm-induced oscillating wake is formed by the slow propagating, near-inertial gravity baroclinic waves, while the fast propagating barotropic waves produce a broad area of convergent depth-averaged currents with no discernible wake; the latter is determined entirely by the wind stress curl, with negligible effects due to the earth's rotation and ocean stratification [10]. The work of [11] confirmed that the mixed layer dynamics is associated with shear-induced entrainment mixing, and forced by near-inertial motions up to the third day after the passage of the storm.

Tropical storms also exert a strong influence on the oceanic chlorophyll *a* field and primary production in the ocean. Another important phenomenon triggered and enhanced by tropical cyclones is the phytoplankton bloom accompanied by nutrient pumping into the oligotrophic surface layer. Concentration of nitrate, phosphate, and chlorophyll *a* is observed to significantly increase after the occurrence of cyclonic disturbances [12]. In [13], after Typhoon Fengwong and Typhoon Sinlaku passed over the southern East China Sea in 2008, the in-situ particulate organic carbon flux was observed to experience a significant increase of about 1.7-fold and 1.5-fold, respectively, compared to the recorded (140–180 mg cm^{-2} d^{-1}) pre-typhoon concentration. The phytoplankton population growth was constrained by the light limitation and the grazing pressure. This increase of the surface chlorophyll *a* concentration might last 2–3 weeks before relaxing to pre-typhoon levels [14]. Because of the limitations imposed by in-situ point observations from ships or moored buoys along a typhoon's track, studies of the associated biological responses have become to more and more depend on the satellite observations. In the work of [15], satellite data are used as new evidence to quantify the contribution of tropical cyclones to enhance the ocean primary production. It was found that the peak of the chlorophyll *a* concentration enhancement tended to occur several days after the sea surface temperature cooling had achieved the maximum amplitude, after the typhoon's passage, see [16, 17]. The extended region and concentration of primary production bloom tend to vary in response to the translation speed and intensity of typhoon. Weak slow-moving typhoons can cause enhanced concentrations of chlorophyll *a*, while strong fast-moving typhoon tends to cause more intense phytoplankton blooms over a larger geographic range, in the wake of the storm as discussed in [18]. Upwelling induced by tropical storms is within the region where the typhoon-induced enhancement of the chlorophyll *a* concentration occurs; please see [19]. The pre-existing cyclonic eddy is capable of strengthening the typhoon-induced nutrient pumping. However, the extent to which that upwelling contributes in the phytoplankton bloom is not clear yet. The development of biological models is helpful to investigate the relevant dynamic process of the nitrogen and carbon cycle in the ocean [20–22].

The ocean temperature cooling and mixed layer deepening caused by tropical storms, usually are underestimated in three-dimensional ocean model simulations, because of the insufficient mixing [23, 24]. Improvements of the mixing estimates in the simulation leading to more realistic and reliable simulations may solve this problem. After the potential energy injected into the mixed layer by tropical storms, wave dispersion spreads energy in both the vertical and meridional directions from the mixed layer [1]. Surface waves have been measured and simulated and shown to play a certain role on enhancing the turbulence in the subsurface layer [25, 26]. In the work of [24], a surface wave model was coupled to a three-dimensional ocean current model (POM), and some related estimates for transfer of momentum and wave energy are derived, assuming linear theory for vertically dependent wave motions. The effect of wave breaking on the simulation of sea surface temperatures and surface boundary layer deepening was investigated in [27]. A wave-induced vertical viscosity (B_V) term, as a function of wave number spectrum, was developed and applied in a global ocean circulation model [28, 29]. This surface wave-induced vertical viscosity term also has a key role on the simulation of sea surface temperature cooling and the mixed layer deepening after typhoon's passage.

In our study, a biological model is coupled to the three-dimensional Regional Ocean Modeling System (ROMS) to investigate both the physical and biological process of Upper Ocean in response to Typhoon Cimaron (2006) which occurred in the South China Sea. Typhoon Cimaron formed over the Pacific Ocean east of the Philippines on October 28, 2006, and then propagated westward, entering the South China Sea on October 30. Thereafter, Cimaron moved to the northwest part of the South China Sea and remained quasistationary during November 1–2, moving southwest on November 3, and finally dissipating near the Vietnamese coast on November 7. Both the mixed layer deepening and sea temperature cooling are underestimated in the model simulation, in comparison to reliable estimates of mixed layer deepened (about 104 m) on November 3 using a one-dimensional remote sensing model [30]. Therefore the wave-induced mixing term B_V is incorporated into ROMS model to improve the accuracy of the simulation.

2. Materials and methods

ROMS is utilized to simulate the processes of Typhoon Cimaron, which influenced the South China Sea from October 29 to November 7, 2006. The model region covers the whole South China Sea including 0°–30°N, 99°–130°E. There are 220 × 100 orthogonal curvilinear grids with the horizontal resolution varying at Δx (5.5–40 km) and Δy (3.6–37 km) and the minimum and maximum depths are 5 m and 5000 m, respectively (**Figure 1**). There are 80 layers in the vertical direction using the *s*-coordinate formulation. The western boundary is closed and the other three open boundaries are defined by the radiation boundary condition.

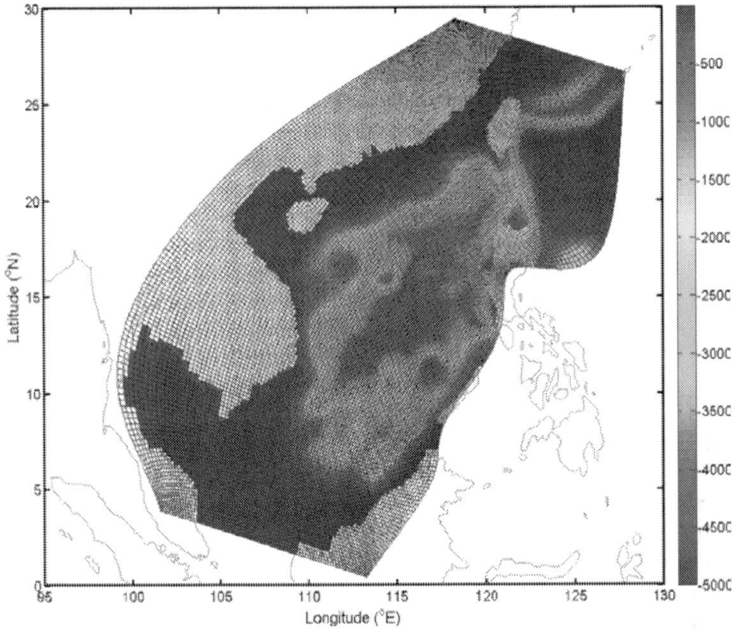

Figure 1. Model domain and grids, as well as bathymetry.

The initial temperature and salinity conditions are taken from the 1/4° grid climatological temperature and salinity analyses of October from WOA01 to represent the pre-typhoon conditions in the South China Sea, which was discussed by Carton and Giese [31]. The climatological monthly data have 24 standard levels with depths varying from 0 to 1500 m and the seasonal data have 33 standard levels with depths from 0 to 5500 m. As the maximum depth in the model is 5000 m, the climatological monthly data are applied in the upper 1500 m and the climatological seasonal data (autumn) are applied from 1500 to 5000 m. The initial current velocity is set to zero in this study.

The lateral boundary conditions for temperature, salinity, sea level and current velocities are obtained from the 5-day averages from the global simulations of the Simple Ocean Data Assimilation (SODA) dataset with horizontal resolution of 0.5° × 0.5° and 40 vertical layers [31]. The Kuroshio Current transport can be identified on the eastern boundary. The tidal amplitudes and phases used in this model are obtained from the TPXO Global Inverse solution database [32] with eight primary tide constituents (M2, S2, N2, K2, K1, O1, P1, and Q1) and two long-period constituents (Mf and Mm).

The daily wind stresses are obtained from QuikSCAT satellite data. The effect of heat flux is considered as the surface boundary for momentum, although the heat flux can be neglected under the extreme meteorological phenomena such as tropical cyclones. The daily heat fluxes are obtained from the Objectively Analyzed air-sea Fluxes (OAFlux) project which is an ongoing research and development project for global air-sea fluxes [33]. To present the climatological heating or cooling trends, we computed the surface boundary conditions for temperature and salinity involving relaxation to its observed values. Corrections are made for the net surface heat flux in the model simulations, as discussed in [34].

For the biological model, the surface chlorophyll *a* field data are estimated from the SeaWiFS climatological seasonal data, and the nitrate (NO_3) and oxygen are estimated from the monthly climatological database of WOA09 [35, 36]. The vertical structure of chlorophyll *a* is extrapolated from the surface chlorophyll *a* field using the Morel and Berthon [37] parameterization method. High concentrations of chlorophyll *a* are mostly distributed in coastal regions along the coastline of southern China and the Vietnam marginal continent, with values of over 1 mg cm^{-3}. By comparison, in the deep sea area, the concentration is less than 0.1 mg cm^{-3}. In most areas of the South China Sea, the NO_3 content is much smaller, less than 0.1 mmol Nm^{-3} in the deep ocean surface layer. The concentration of NO_3 increases with ocean depth. In the deep sea, with depths in excess of 800 m, the concentration of NO_3 can reach 40 mmol Nm^{-3}.

Both the total inorganic carbon (TIC) and total alkalinity are obtained from the Carbon Dioxide Information Analysis Center. The detailed data information is given in work et al. [38]. The climatological seasonal dataset is used to generate the initial and boundary conditions. Ammonium (NH_4), large and small detritus, and N-concentration are taken from the NO_3 estimates and multiplied by the respective ratios of 18/62, 0.1 and 0.1, respectively, while phytoplankton, zooplankton, large and small detritus, and C-concentration are taken from the chlorophyll *a* concentration multiplied by the respective ratios of 0.5, 0.2, 0.1, and 0.1, respectively.

3. Model validation

Two Optimally Interpolated (OI) SST daily products (including microwave plus infrared (MWIR) OI SSTs and microwave only TMI AMSRE SSTs) are used in comparisons with the simulated SSTs to validate the model accuracy. MWIR OI SST product is at 9 km resolution, while TMI AMSRE SST product is at 25 km resolution. The validation statistical parameters contain mean error (ME), mean absolute error (MAE), root-mean-square (RMS), and correlation coefficient (R). The formulas to calculate these statistical parameters are presented in [39].

The simulated SSTs (**Figure 2**) are compared with the satellite observations from October 30 to November 6. The validation region is set from 99°E to 120°E and from 0°N to 26°N. For validation of the simulation, the statistical parameters are displayed in **Table 1** for MWIR OI and **Table 2** for TMI AMSRE. The MEs of the simulated SSTs are less than 0.12°C, compared with the MWIR SSTs; for TMI AMSRE SSTs, the MEs are within 0.14–0.24°C. Negative signs indicate that the modelled SSTs are less than the observed SSTs. The MAEs are within the range from 0.4°C to 0.6°C, and the RMS errors are less than 0.9°C. The correlation coefficients between the MWIR OI SST and simulated SST are over 87% from October 30 to November 6, while the correlation coefficients validated with TMI AMSRE SSTs are over 84%. The high values of the correlation coefficients indicate that the simulated SSTs are within a reasonable range. Therefore, we have demonstrated that ROMS can generally reproduce the processes of Typhoon Cimaron in the South China Sea.

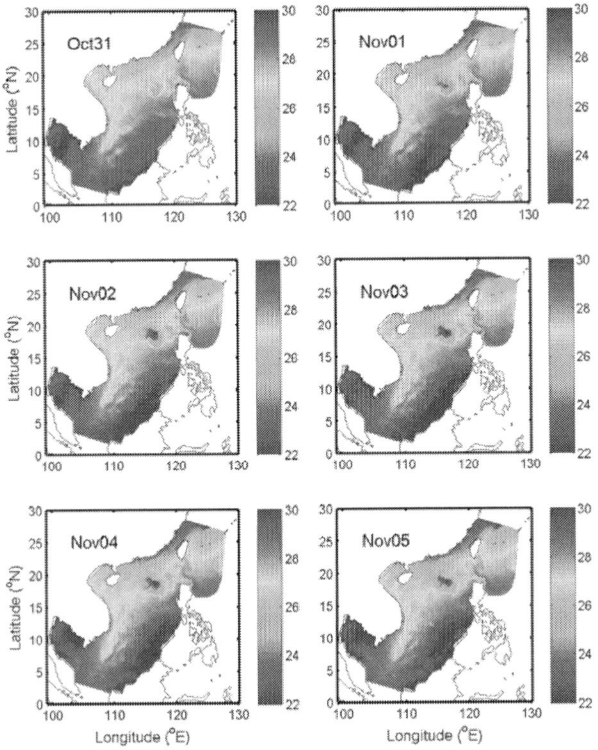

Figure 2. The simulated SSTs (unit: °C) from October 31 to November 5, 2006.

	Oct 30	Oct 31	Nov 1	Nov 2	Nov 3	Nov 4	Nov 5	Nov 6
ME (°C)	−0.12	−0.12	−0.09	−0.04	−0.01	−0.00	−0.09	−0.10
MAE (°C)	0.42	0.45	0.50	0.54	0.58	0.56	0.55	0.58
RMS (°C)	0.56	0.61	0.69	0.76	0.88	0.81	0.78	0.80
R	0.89	0.88	0.88	0.89	0.87	0.88	0.88	0.88

Table 1. Statistics of the simulated SSTs versus MWIR OI SSTs from October 30 to November 6, 2006.

http://dx.doi.org/10.5772/64099

	Oct 30	Oct 31	Nov 1	Nov 2	Nov 3	Nov 4	Nov 5	Nov 6
ME (°C)	−0.23	−0.24	−0.21	−0.14	−0.14	−0.17	−0.21	−0.16
MAE (°C)	0.41	0.44	0.48	0.53	0.53	0.49	0.46	0.46
RMS (°C)	0.53	0.58	0.71	0.77	0.78	0.71	0.68	0.67
R	0.87	0.86	0.84	0.87	0.88	0.89	0.89	0.90

Table 2. Statistics of the simulated SSTs versus TMI AMSRE SSTs from October 30 to November 6, 2006.

4. Results

4.1. Ocean temperatures

The ocean surface temperature distributions from October 31 to November 5, 2006 are shown in **Figure 2**. Cimaron entered the South China Sea on October 30, and the simulation of the typhoon-induced wake suggests that the apparent temperature depression started on October 31. The wake of the typhoon occurred within the region from 15°N to 20°N and from 114°E to 119°E. **Figure 3** shows the time series (daily) of the maximum SST decreasing in the typhoon-induced wake, located at 18.51°N and 116.45°E, relative to the pre-typhoon conditions on October 28. Although bias exists between these two satellite SST observations, both satellite SSTs display the same trend in their variations in the SST cooling amplitude in the wake. The

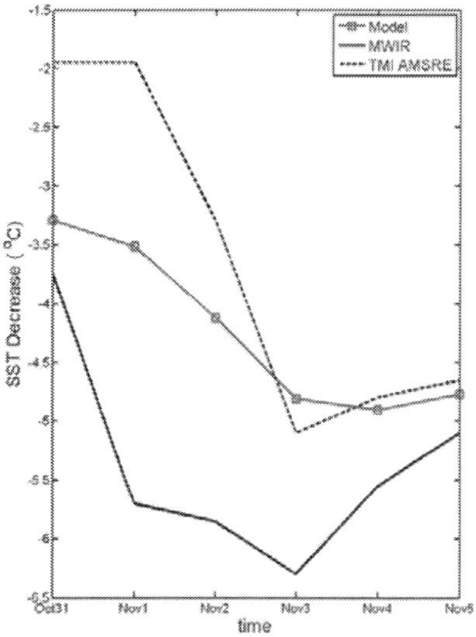

Figure 3. Maximum SST decrease in the typhoon wake from October 31 to November 5, 2006 (the red solid line with squares denotes the model results; the dash line denotes TMI AMSRE satellite observations; the solid line represents MWIR OI satellite observations).

observed SST decreased to an extremely low value on November 3, by which time the ocean was in the forcing stage. Starting from November 3, SSTs had begun to increase, which defines the beginning of the relaxation stage towards a new equilibrium state, with the injection of potential vorticity into the wake by the typhoon winds [1]. Comparing model results to these two satellite SSTs which show that the maximum amplitude of the SST cooling appeared on November 3, with values of 5.1°C for TMI AMSRE and 6.3°C for MWIR OI, the simulated maximum SST cooling was 4.8°C on November 4 (**Figure 3**). Thus, the relaxation of surface temperature in the wake after the typhoon's passage is clearly underestimated in our simulations.

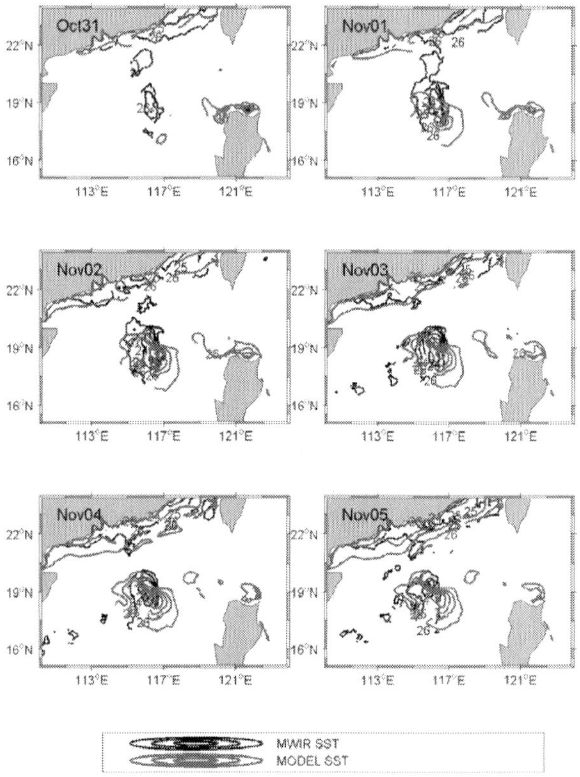

Figure 4. Isotherms of SST in the typhoon wake (unit: °C) from October 31 to November 5, 2006 (with the black lines showing the MWIR OI SSTs and blue lines showing the simulated SSTs).

Figure 4 shows the comparison of the daily surface isotherms between MWIR OI observations and the model simulations from October 31 to November 5. The isothermal lines ranged from 22°C to 26°C with 1°C intervals. The distributions of the simulated surface isotherms in the wake area are quite consistent with the MWIR OI SSTs, which further demonstrate that the temperature simulations related to Typhoon Cimaron are well reproduced. The southward shift of the location of the maximum SST cooling in the simulation results compared to the satellite observations indicates another common issue during the forecasting of extreme weather conditions, which is the lack of high-accuracy wind forcing observations. The simulated maximum mixed layer depth was about 53.2 m, located at (18.51°N, 116.44°E) on November 3, 2006 which is an underestimate compared with the maximum deepening of 104.5 m at (19.50°N, 116.26°E) estimated in work [30]. Underestimation of the mixed layer depth and SST cooling is a common problem in the numerical ocean model simulations because of insufficient mixing. To solve this problem, a parameterization of wave-induced mixing is added to the model to improve the mixing estimates in this chapter (see details in Section 5).

4.2. Ocean currents

When Typhoon Cimaron entered the South China Sea on October 30, the upper ocean's response was almost instantaneous. Typhoon-induced cyclonic currents caused divergence of upper ocean water over the surface areas by tens of kilometres (in scale) with cold water that is upwelled from the deeper ocean, accounting for the formation of a cold-core eddy. The cyclonic currents flowed in a roughly circular motion around the mesoscale cold eddy, with the maximum velocity reaching 2.5 m s^{-1} at location (18.44°N, 115.3°E) on October 31. An alongshore current flowing through the Taiwan Strait into the South China Sea, driven by the northeast monsoon winds, was strengthened by the typhoon forcing on October 31 and November 1. The intrusion of Kuroshio Current meandering towards the South China Sea through the Luzon Strait was strengthened too, with the current velocity reaching 1.6 m s^{-1} north of the Philippines Islands on November 1.

A very large amount of potential and kinetic energy is injected into the ocean surface layer from the strong typhoon winds during the typhoon generation and development process. The power of the injected energy and the vorticity can be calculated using the following formulae:

$$W = \vec{F} \cdot \vec{U} = \tau_x u + \tau_y v \tag{1}$$

$$\zeta = \frac{\partial v}{\partial x} - \frac{\partial u}{\partial y} \tag{2}$$

where W represents the power in the units of W m^{-2}. Here, $\vec{F}$ represents the wind stress with (τ_x, τ_y) the two components corresponding to the (x, y) coordinates, respectively. $\vec{U}$(u, v) is the simulated surface current velocity and ζ represents the vorticity.

Strong cyclonic currents caused by Typhoon Cimaron lasted for several days after typhoon's passage. Strengthened by the long-lasting intense typhoon wind forcing, the cyclonic eddy generated by the cyclonic circulation reached a maximum positive vorticity of 3.56 × 10^{-5} s^{-1} at the location (116.8°E, 18.3°N) on October 30, and continued intensifying to 5.36 × 10^{-5} s^{-1} on October 31 and 5.07 × 10^{-5} s^{-1} on November 1 (**Figure 5**). **Figure 6** shows estimates for the power of the injected oceanic kinetic energy from October 30 to November 4. The power input from the wind forcing was mostly located on the right side of the typhoon's track with maxima of 1.45 W m^{-2} on October 31, and 1.43 W m^{-2} on November 1 and 0.54 W m^{-2} on November 2. The power input was low in the typhoon's eye area where the winds were very weak. The initial oceanic state was changed under the effect of the strong typhoon winds. On October 30, there is a large region with negative kinetic energy injected by the typhoon, appearing on the right side of the typhoon's track; this suggests that the oceanic kinetic energy was decreased by the typhoon. Thus, the intense typhoon winds changed the original pre-typhoon sea state conditions. The maximum decrease in the oceanic kinetic energy in the wake of the typhoon was 0.83 W m^{-2}. On October 31, the power reached a maximum value of 2.3 W m^{-2}, and on November 1, this value was 2.0 W m^{-2}.

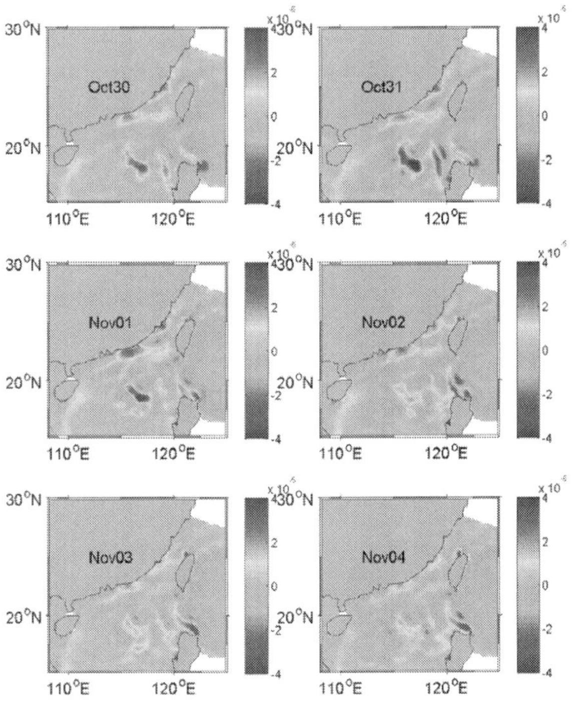

Figure 5. Surface vorticity from October 30 to November 4, 2006.

From November 2 onwards, the injected energy into the wake region of the typhoon showed the decreasing tendency pattern. However, the vorticity of the current induced by typhoon persisted for a long time after typhoon's passage. The study presented here shows that the energy input induced by the typhoon winds was responsible for the ocean-enhanced mixing processes.

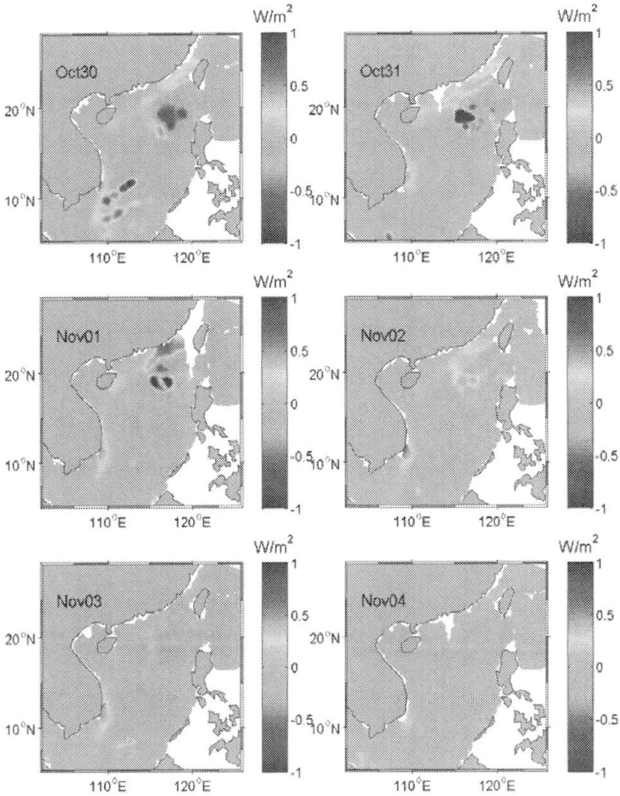

Figure 6. The power input to the ocean surface by the typhoon winds from October 30 to November 4, 2006.

4.3. Biological results

The monthly chlorophyll *a* concentration gridded datasets are obtained from the SeaWiFS observations, with a horizontal spatial resolution of 0.0417° × 0.0417°. **Figure 7** shows the monthly distributions of surface chlorophyll *a* concentration on October 16 and November 16, 2006 in the South China Sea. These two satellite images represent the pre- and post-typhoon situations for the primary production; it is evident that a phytoplankton bloom area exists around the location of the typhoon's wake after it had passed. Although the concentration of the chlorophyll *a* achieved high values, in excess of 4 mg cm^{-3} along the coast, the averaged concentration was about 0.09 mg cm^{-3} in the area that became the wake (the black box in **Figure 7**) before the appearance of Typhoon Cimaron. However, by November 16, the concentration in this area reached 0.85 mg cm^{-3}, which was much higher than the normal expected conditions, indicating that the maximum increase of chlorophyll *a* concentration was about 0.75 mg cm^{-3} at the location (17.02°N, 115.57°E) in the wake area. While the monthly dataset may be interpolated in both spatial and temporal dimensions, the detailed processes related to the development of the phytoplankton bloom triggered by Typhoon Cimaron are still not clear. The application of the biological numerical model can provide more details for the study of the subsurface water layers and can complement the limitations of the satellite remote sensing.

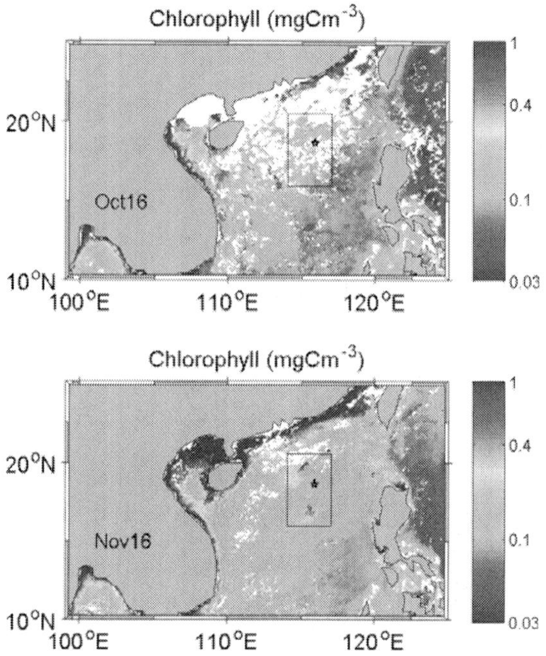

Figure 7. Chlorophyll *a* concentration (unit: mg cm^{-3}) from SeaWiFS on October 16 (upper panel) and November 16 (lower panel), 2006 (black box represents the areas in the wake of the typhoon; the pentagram marks the location used for calculation in **Figure 9**).

Figure 8 shows the simulated daily maximum surface concentrations of chlorophyll *a* and nitrate in the typhoon's wake area through October 28 to November 30. In the pre-typhoon condition, the nitrate concentration maintained a stable value of 0.03 mmol Nm^{-3}. During November 1–3, Cimaron lingered in locations that were quasistationary and caused strong mixing in the wave areas. Thus the nitrate concentration largely increased in the ocean surface layer in the wake, reaching a maximum of 1.24 mmol Nm^{-3} on November 3. By November 3, Cimaron had passed from this (wake) region. The surface nitrate concentration remained high, in excess of 1.1 mmol Nm^{-3} for an additional 3 days, from November 3–5, and then decreased to a rather stable level of 0.1 mmol Nm^{-3} from November 11 onwards.

Compared to the quick response of nitrate to Typhoon Cimaron, the response of the phytoplankton is a relatively slow process. The chlorophyll *a* concentration remained at the pretyphoon level of about 0.06 mg cm^{-3} for some time beyond November 3. Although the mixed layer depth deepened to its maximum on November 3, the chlorophyll *a* concentration was still increasing at that time, having increased slightly over the previous 2 days, as triggered by the upwelling induced by the approaching storm. Thereafter, the chlorophyll *a* concentration increased rapidly in early November, reaching a rate of 0.5 mg cm^{-3} d^{-1}, and attaining a maximum concentration of 1.76 mg cm^{-3} on November 7. The phytoplankton blooms occurred 5 days after Typhoon Cimaron's passage. The chlorophyll *a* concentration began to decrease from November 8 onwards, returning to a quasistable level of 0.3 mg cm^{-3} by November 18. The maximum concentration of chlorophyll *a* on November 16 was simulated at the value of 0.65 mg cm^{-3}, which was about 0.2 mg cm^{-3} less than the satellite observations. The surface ocean was restored to an equilibrium state again by about 10–20 days after the interruption that Cimaron introduced. Moreover, the concentrations of both nitrate and chlorophyll *a* in the resulting re-equilibrium of the ocean state are higher than those of the former pre-typhoon state.

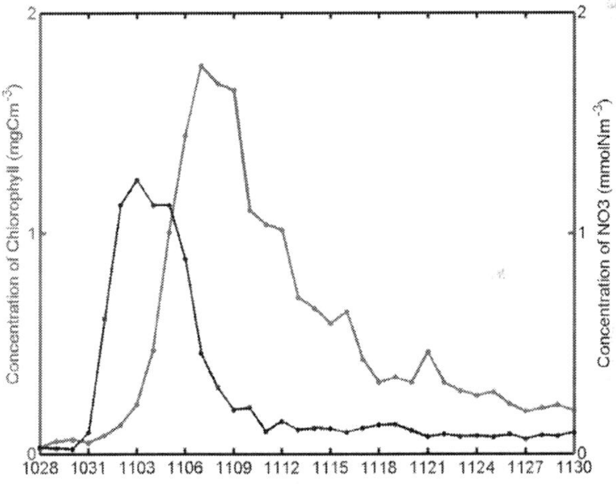

Figure 8. Simulated maximum concentrations of nitrate and chlorophyll *a* in the typhoon wake from October 28 to November 30, 2006.

The vertical profiles of the density, both of chlorophyll *a* and nitrate concentrations at the location (18.66°N, 115.89°E), as shown in **Figure 7**, in the typhoon wake are investigated with respect to the underwater impacts of Cimaron, comparing the pre- and post-typhoon profiles through October 28 to November 30 in **Figure 9**. Before the typhoon, the nutrient and phytoplankton in the surface layer are both at low concentrations, as the surface waters received strong light irradiation, which is not conducive to the growth and reproduction of the phytoplankton. Phytoplankton populations grow and reproduce mostly in the euphotic zone. The depth of the euphotic zone can be estimated from the chlorophyll *a* concentration of the surface layer, based on the assumption of Case-I waters; the equation is shown in [40]. After the typhoon's passage, cyclonic eddies caused by Typhoon Cimaron exhibited the upward doming of isopycnals from October 31 and the isopycnals were uplifted with high nutrient concentrations into the euphotic zone, which furthermore, had a positive influence on the photosynthetic performance. The chlorophyll *a* concentration in the surface layer increased and reached about 0.237 mg cm^{-3} on November 7 in **Figure 9**. Both the profiles of nitrate and chlorophyll *a* are significantly elevated after the typhoon's passage. The euphotic zone depth was estimated as 87.0 m before typhoon and 36.2 m after typhoon in the wake, with respect to the chlorophyll *a* concentrations obtained from the satellite for pre- and post-typhoon Cimaron. The euphotic zone was uplifted by 50.0 m in the wake of Typhoon Cimaron.

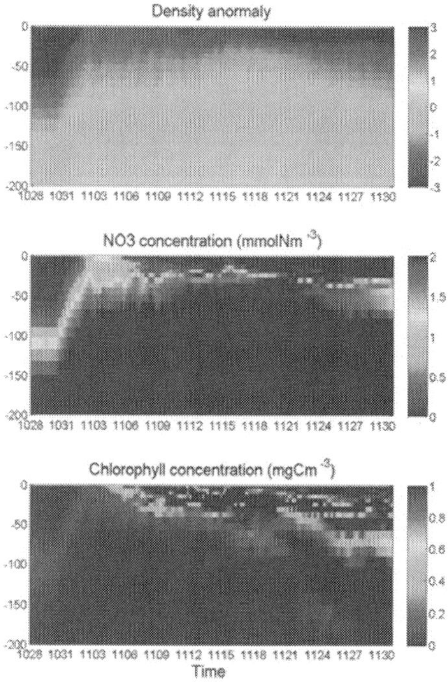

Figure 9. Profiles of chlorophyll *a* and nitrate concentration at the location (18.66°N, 115.89°E), shown in **Figure 7** in the typhoon wake from October 28 to November 30, 2006.

5. Discussion: effect of the wave-induced mixing

The mixed layer deepening induced by Typhoon Cimaron is underestimated in the three-dimensional ocean model simulations, which is a common situation in ocean model simulations. To strengthen the insufficient mixing, in our chapter, we incorporated the wave-induced mixing term, B_V into ROMS to investigate the effect of B_V on the mixed layer deepening and ocean surface temperature cooling caused by Typhoon Cimaron.

The wave-induced mixing term B_V, is added into ROMS, as part of the vertical kinematic viscosity, as expressed by Qiao et al. [29]

$$B_V = \alpha \iint_{\vec{k}} E(\vec{k})\exp(2kz)d\vec{k}\frac{\partial}{\partial z}\left[\iint_{\vec{k}} \omega^2 E(\vec{k})\exp(2kz)d\vec{k}\right]^{1/2} \quad (3)$$

where ω is the wave angular frequency, z is the vertical coordinate axis downward positive with $z = 0$ at the surface, k is the wave number, and $E(\vec{k})$ represents the wave number spectrum including both wind wave and swell waves. B_V can be calculated by the wave model. In this study, the wave-induced mixing term is directly derived from the Key Laboratory of Marine Science and Numerical Modeling (MASNUM) wave model [28] and is added into the vertical viscosity and diffusivity term of the K-profile parameterization (KPP) mixing scheme which is applied in ROMS. Because the wave mixing is dominant in the upper ocean layer, the B_V term is confined to the range from 1000 m depth up to the surface. The weighting coefficient was set to 0.1 in our study, as suggested by Wang et al. [41]. In the wake of the typhoon, the vertical mixing was strengthened with the vertical viscosity coefficient of 0.1 $m^2 s^{-1}$ from October 31 to November 2.

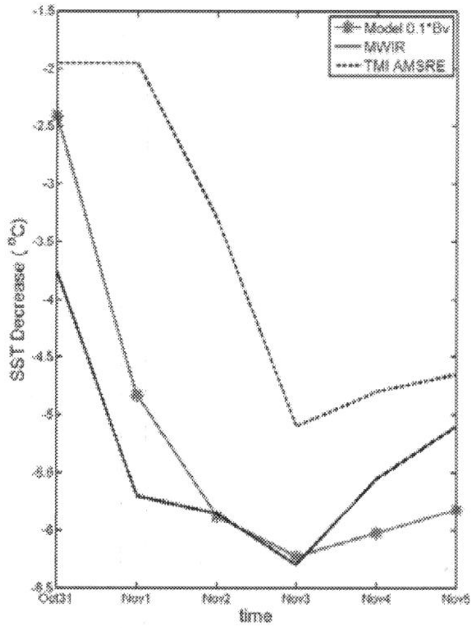

Figure 10. Maximum SST decrease in the wake from October 31 to November 5, 2006 (red solid line with asterisks denotes the model results; the dash line denotes TMI AMSRE satellite observations; the solid line represents MWIR satellite observations).

Figure 10 shows the maximum SST decrease in the wake of the typhoon, comparing the simulations by adding B_V with the two sets of satellite observations. Under the effect of the strengthened mixing estimates, SST in the wake reached the lowest temperature on November 3 with a value that is consistent with both sets of satellite observations. The maximum SST decreases on November 2 and November 3, respectively, relative to the pre-typhoon conditions on October 28, which were 5.9°C and 6.2°C, which are close to the MWIR observations of 5.8°C and 6.3°C. Compared to the maximum temperature decreases without the B_V term of 4.1°C on November 2 and 4.8°C on November 3 in the typhoon's wake, the wave-induced mixing can improve the SST cooling by 1.7°C on November 2 and 1.4°C on November 3. This is with a weighting coefficient of 0.1. The associated mixed layer deepening was increased by 30 m on November 3.

6. Conclusions

A three-dimensional simulation of the upper ocean in response to Typhoon Cimaron is investigated in this study, including both the physical and biological processes. The validation of SST was compared with two satellite observations, TMI AMSRE and MWIR OI SSTs, from October 30 to November 6. High correlation (over 84%) and low bias (between 0.4°C and 0.6°C) show that ROMS can reproduce the process of upper ocean response to Typhoon Cimaron quite well. Detailed analysis indicates that the surface cooling is underestimated due to the insufficient mixing in the ROMS model. To solve this problem, the wave-induced mixing with a certain weighting coefficient was introduced into the KPP mixing scheme to improve the simulation of SST cooling. Values up to 6.2°C are obtained, which is close to the observed MWIR cooling estimate of 6.3°C on November 3, whereas the ROMS simulation without the wave-induced mixing gives an underestimated cooling of 4.8°C. The simulation accuracy is enhanced by adding the wave-induced mixing, which increases the SST cooling by 1.4°C and deepens the mixed layer by 30 m in the wake of typhoon.

A strong mesoscale ocean eddy, as characterized by the cyclonic currents, was caused by Typhoon Cimaron in the South China Sea. The water within the eddy diverged over surface areas on a scale of tens of kilometres. Under the divergent condition, cold nutrient-rich water upwelled from deeper waters. The positive vorticity kept a high value over $5.0 \times 10^{-5}\ s^{-1}$ on October 31 and November 1. Moreover, the concentration of nitrate in the surface wake area increased to a maximum during these 2 days, which indicates that upwelling played a key role on the phytoplankton blooming after typhoon's passage. The simulated maximum concentration of chlorophyll *a* in the wake increased from a pre-typhoon value of 0.1 mg cm^{-3} to a post-typhoon value of 0.65 mg cm^{-3} on November 16, which is close to the satellite observation of 0.85 mg cm^{-3} on November 16. The euphotic zone was uplifted by 50.0 m after Typhoon Cimaron's passage. Thereafter, the ocean restored to a new equilibrium state with higher concentrations of chlorophyll *a* and nitrate than those existing in the pre-equilibrium state in the wake area.

Acknowledgements

We thankfully acknowledge funding support from the General Research Fund of Hong Kong Research Grants Council (RGC) under Grants CUHK 402912 and 403113, the Hong Kong Innovation and Technology Fund under the Grants of ITS/321/13, the direct grants of the Chinese University of Hong Kong, the National Natural Science Foundation of China under project 41376035, the Marine Environmental Observation Prediction and Response Network (MEOPAR), DFO's Aquatic Climate Change Adaptation Service (ACCASP), the PERD, and the Canadian Panel on Energy Research and Development.

Appendices

Appendix A – Acronyms

B_V	Wave-induced mixing
K1	Lunisolar declinational diurnal tidal constituent
K2	Lunisolar declinational semidiurnal tidal constituent
KPP	K-profile parameterization
M2	Principal lunar semidiurnal tidal constituent
MAE	Mean absolute error
MASNUM	Key Laboratory of Marine Science and Numerical Modeling
ME	Mean error
Mf	Lunar fortnightly long-period tidal constituent
Mm	Lunar monthly long-period tidal constituent
MWIR OI	Microwave infrared optimally interpolated
N2	Larger lunar elliptic semidiurnal tidal constituent
NH_4	Ammonium
NO_3	Nitrate
R	Correlation coefficient
RMS	Root-mean-square
ROMS	Regional Ocean Modeling System
S2	Principal solar semidiurnal tidal constituent
SeaWIFS	Sea-Viewing Wide Field-of-View Sensor
SSTs	Sea surface temperatures
SODA	Simple Ocean Data Assimilation dataset

TIC	Total inorganic carbon
TMI AMSRE	Tropical Rainfall Measuring Mission (TRMM) Microwave Imager and Advanced Microwave Scanning Radiometer
TPXO	The OSU TOPEX/Poseidon Global Inverse Solution
O1	Principal lunar diurnal tidal constituent
OAFlux	Objectively Analyzed air-sea Fluxes
P1	Principal solar diurnal tidal constituent
POM	Princeton Ocean Model
Q1	Major lunar elliptical diurnal tidal constituent
QuikSCAT	NASA Quick Scatterometer Earth observation satellite
WOA01	World Ocean Atlas 2001
WOA09	World Ocean Atlas 2009

Author details

Yujuan Sun[1,2,3], Jiayi Pan[1,4,5*] and William Perrie[2,3]

*Address all correspondence to: panj@cuhk.edu.hk

1 Institute of Space and Earth Information Science, The Chinese University of Hong Kong, Hong Kong, China

2 Department of Engineering Mathematics and Internetworking, Dalhousie University, Halifax, Nova Scotia, Canada

3 Fisheries and Oceans Canada, Bedford Institute of Oceanography, Dartmouth, Nova Scotia, Canada

4 Shenzhen Research Institute, The Chinese University of Hong Kong, Shenzhen, China

5 School of Marine Science, Nanjing University of Information Science and Technology, Nanjing, Jiangsu, China

References

[1] Gill AE: On the behavior of internal waves in the wakes of storms. J. Phys. Oceanogr. 1984; 14: 1129–1151.

[2] Price JF, Sanford TB, Forristall GZ: Forced stage response to a moving hurricane. J. Phys. Oceanogr. 1994; 24: 233–260.

[3] Leipper DF: Observed ocean conditions and Hurricane Hilda 1964. J. Atmos. Sci. 1967; 24: 182–196.

[4] Bolin B: The adjustment of a non-balanced velocity field towards geostrophic equilibrium in a stratified fluid. Tellus. 1953; 5(3): 373–385.

[5] Pollard RT: On the generation by winds of inertial waves in the ocean. Deep-Sea Res.. 1970; 17(4): 795–812. DOI: 10.1016/0011-7471(70)90042-2.

[6] Veronis G: Partition of energy between geostrophic and non-geostrophic oceanic motions. Deep-Sea Res.. 1956; 3(3): 157–177. DOI: 10.1016/0146-6313(56)90001-6.

[7] Pollard RT, Millard Jr RC: Comparison between observed and simulated wind-generated inertial oscillations. Deep-Sea Res.. 1970; 17(4): 813–821. DOI: 10.1016/0011-7471(70)90043-4.

[8] Blumen W, Wu R: Geostrophic adjustment: frontogenesis and energy conversion. J. Phys. Oceanogr. 1995; 25: 428–438.

[9] Geisler JE: Linear theory of the response of a two layer ocean to moving hurricane. Geophys. Fluid Dyn. 1970; 1: 249–272. DOI: 10.1080/03091927009365774.

[10] Ginis I, Sutyrin G: Hurricane-generated depth-averaged currents and sea surface elevation. J. Phys. Oceanogr. 1995; 25: 1218–1242.

[11] Black PG: The 3D oceanic mixed layer response to Hurricane Gilbert. J. Phys. Oceanogr. 2000; 30: 1407–1429.

[12] Chang J, Chung CC, Gong GC: Influences of cyclones on chlorophyll *a* concentration and Synechococcus abundance in a subtropical western Pacific coastal ecosystem. Mar. Ecol. Prog. Ser. 1996; 140: 199–205.

[13] Hung CC, Gong GC, Chou WC, Chung CC, Lee MA, Chang Y, Chen HY, Huang SJ, Yang Y, Yang WR, Chung WC, Li SL, Laws E: The effect of typhoon on particulate organic carbon flux in the southern East China Sea. Biogeosciences. 2010; 7: 3007–3018.

[14] Babin SM, Carton JA, Dickey TD, Wiggert JD: Satellite evidence of hurricane-induced phytoplankton blooms. J. Geophys. Res. 2004; 109: C03043. DOI: 10.1029/2003JC001938.

[15] Lin I, Liu W, Wu CC, Wong TF, Hu C, Chen Z, Liang WD, Yang Y, Liu KK: New evidence for enhanced ocean primary production triggered by tropical cyclone. Geophys. Res. Lett. 2003; 30(13). DOI: 10.1029/2003GL017141.

[16] Zheng G, Tang D: Offshore and nearshore chlorophyll increases induced by typhoon winds and subsequent terrestrial rainwater runoff. Mar. Ecol. Prog. Ser. 2007; 333: 61–74.

[17] Shang S, Li L, Sun F, Wu J, Hu C, Chen D, Ning X, Qiu Y, Zhang C, Shang S: Changes of temperature and bio-optical properties in the South China Sea in response to

Typhoon Lingling 2001. Geophys. Res. Lett. 2008; 35: L10602. DOI: 10.1029/2008GL033502.

[18] Zhao H, Tang D, Wang Y: Comparison of phytoplankton blooms triggered by two typhoons with different intensities and translation speeds in the South China Sea. Mar. Ecol. Prog. Ser. 2008; 365: 57–65.

[19] Chang Y, Liao HT, Lee MA, Chan JW, Shieh WJ, Lee KT, Wang GH, Lan YC: Multisatellite observation on upwelling after the passage of Typhoon Hai-Tang in the southern East China Sea. Geophys. Res. Lett. 2008; 35: L03612. DOI: 10.1029/2007GL03285.

[20] Franks PJS, Wroblewski JS, Flierl GR: Behavior of a simple plankton model with food-level acclimation by herbivores. Mar. Biol. 1986; 91: 121–129.

[21] Fennel K, Wilkin J, Levin J, Moisan J, O'Reilly J, Haidvogel D: Nitrogen cycling in the Middle Atlantic Bight: results from a three-dimensional model and implications for the North Atlantic nitrogen budget. Global Biogeochem. 2006; 20: GB3007. DOI: 10.1029/2005GB002456.

[22] Powell TM, Lewis CVW, Curchitser EN, Haidvogel DB, Hermann AJ, Dobbins EL: Results from a three-dimensional, nested biological-physical model of the California Current System and comparisons with statistics from satellite imagery. J. Geophys. Res. 2006; 111: C07018. DOI: 10.1029/2004JC002506.

[23] Ezer T: On the seasonal mixed layer simulated by a basin-scale ocean model and the Mellor-Yamada turbulence scheme. J. Geophys. Res. 2000; 105: 16843–16855.

[24] Mellor G: The three-dimensional current and surface wave equations. J. Phys. Oceanogr. 2003; 33: 1978–1989.

[25] Sanford TB, Black PG, Haustein JR, Feeney JW, Forristall GZ, Price JF: Ocean response to a hurricane. Part I: observations. J. Phys. Oceanogr. 1987; 17: 2065–2083.

[26] Craig PD, Banner ML: Modeling wave-enhanced turbulence in the ocean surface layer. J. Phys. Oceanogr. 1994; 24: 2546–2559.

[27] Mellor G, Blumberg A: Wave breaking and ocean surface layer thermal response. J. Phys. Oceanogr. 2004; 34: 693–698.

[28] Yuan Y, Qiao F, Hua F, Wan Z: The development of a coastal circulation numerical model: I. Wave-induced mixing and wave-current interaction. J. Hydrodyn. Ser. A. 1999; 14: 1–8.

[29] Qiao F, Yuan Y, Yang Y, Zheng Q, Xia C, Ma J: Wave-induced mixing in the upper ocean: distribution and application to a global ocean circulation model. Geophys. Res. Lett. 2004; 31: L11303. DOI: 10.1029/2004GL019824.

[30] Pan J, Sun Y: Estimate of ocean mixed-layer deepening after a typhoon passage over the South China Sea by using satellite data. J. Phys. Oceanogr. 2013; 43: 498–506.

[31] Carton JA, Giese BS: A reanalysis of ocean climate using Simple Ocean Data Assimilation (SODA). Mon. Weather Rev. 2008; 136: 2999–3017.

[32] Egbert GD, Bennett AF, Michael MGG: TOPEX/POSEIDON tides estimated using a global inverse model. J. Geophys. Res.. 1994; 99(C12): 24821–24852.

[33] Yu L, Jin X, Weller RA: Multidecade global flux datasets from the Objectively Analyzed Air-sea Fluxes (OAFlux) project: latent and sensible heat fluxes, ocean evaporation, and related surface meteorological variables, Woods Hole Oceanographic Institution, OAFlux Project Technical Report, OA-2008-01. Woods Hole, Massachusetts; 2008. 64 p.

[34] Haney RL: Surface thermal boundary condition for ocean circulation models. J. Phys. Oceanogr. 1971; 1: 241–248.

[35] Garcia HE, Locarnini RA, Boyer TP, Antonov JI, Baranova OK, Zweng MM, Johnson DR. World Ocean Atlas 2009, Volume 3: Dissolved Oxygen, Apparent Oxygen Utilization, and Oxygen Saturation. S. Levitus, Ed. NOAA Atlas NESDIS 70, U.S. Government Printing Office, Washington D.C.; 2010. 344 p.

[36] Garcia HE, Locarnini RA, Boyer TP, Antonov JI, Zweng MM, Baranova OK, Johnson DR: World Ocean Atlas 2009, Volume 4: Nutrients (phosphate, nitrate, silicate). S. Levitus, Ed. NOAA Atlas NESDIS 71, U.S. Government Printing Office, Washington D.C.; 2010. 398 p.

[37] Morel A, Berthon JF: Surface pigments, algal biomass profiles, and potential production of the euphotic layer: relationships reinvestigated in view of remote-sensing applications. Limnol. Oceanogr. 1989; 34: 1545–1562.

[38] Goyet C, Healy RJ, Ryan JP: Global distribution of total inorganic carbon and total alkalinity below the deepest winter mixed layer depths. 2000; NDP-076, http://cdiac.ornl.gov/oceans/ndp_076/.

[39] Sun Y, Qiao F, Wang G, Ying X, Yang Y: Forecast operation and verification of MASNUM surface wave numerical model (in Chinese). Adv. Mar. Sci.. 2009; 27(3): 281–293.

[40] Lee Z, Weidemann A, Kindle J, Arnone R, Carder KL, Davis C: Euphotic zone depth: its derivation and implication to ocean-color remote sensing. J. Geophys. Res. 2007; 112: C03009. DOI: 10.1029/2006JC003802.

[41] Wang Y, Qiao F, Fang G, Wei Z: Application of wave-induced vertical mixing to the K profile parameterization scheme. J. Geophys. Res. 2010; 115: C09014. DOI: 10.1029/2009JC005856.

Chapter 5

Planetary-Scale Low-Level Circulation and the Unique Development of Hurricane Wilma in 2005

Jinwoong Yoo, Robert V. Rohli and Jennifer Collins

Additional information is available at the end of the chapter

http://dx.doi.org/10.5772/64061

Abstract

Large-scale atmospheric and oceanic conditions in the western Atlantic basin were analyzed to understand the unique tropical cyclogenesis (TCG) and intensification mechanism of Hurricane Wilma in 2005, the most intense Atlantic basin tropical cyclone (TC) on record. An analysis of 850 hPa circulations depicted in the National Centers for Environmental Prediction/National Center for Atmospheric Research (NCEP/NCAR) reanalysis data suggests that anomalous development of the 850 hPa circulation pattern triggered by Hurricane Vince (October 8–11, 2005) contributed to the development of a large-scale low-level vortex that preceded Wilma's TCG in the eastern Caribbean. In particular, weakened easterly winds in the central tropical Atlantic assisted the unique large-scale cyclonic circulation over the western Atlantic about a week before Wilma's TCG. The unprecedented rapid intensification of Wilma was investigated considering the interactions between mid-latitude troughs and large-scale low-level circulations as well as anomalously warm SST conditions. The global Weather Research and Forecasting (WRF) model run for Hurricane Wilma suggests that the role of mid-latitude systems in TC activity is more important than previously believed and that every in situ large- or meso-scale vortex and circulation component at least in the immediately neighboring region of TCG seems to have a significant influence on TCG processes.

Keywords: tropical cyclone genesis, Hurricane Wilma, vortices, WRF, low-level circulations

INTECH
open science | open minds

1. Introduction

The importance of large-scale external circulations in tropical cyclogenesis (TCG) seems irrefutable, based on the numerous persuasive previous studies [1–3]. In a composite analysis of tropical cyclones (TCs) that developed from monsoon troughs over the western North Pacific, Briegel and Frank [2] hypothesized that eastward-propagating subtropical troughs poleward of the location of TCG support TCG by providing upper-tropospheric vorticity advection, thereby forcing upper-level divergence and uplift. Briegel and Frank [2] highlighted that successfully developing TCs had 850 hPa southwesterly surges in addition to the monsoonal easterly winds approximately 48–72 hours prior to TCG, potentially triggering the low-level convergence and deep uplift necessary for TCG. Using a global numerical weather prediction model for TCG in the western North Pacific, Chan and Kwok [4] derived a similar conclusion to Briegel and Frank [2] regarding the general synoptic-scale features present before a TCG, specifically the relatively important roles of the low-level trade winds and the southwesterly low-level wind surge prior to TCG. Interestingly, however, Briegel and Frank [2] also found that nongenesis cases have upper-level troughs both to the northwest and to the northeast of the genesis region, which complicates the distinction between TCG-triggering and non-TCG-triggering synoptic settings. Data limitations precluded the specification of the source of the southwesterly surge into the genesis location, other than any preexisting TCs, which only occur in about 34% of all the genesis cases [2].

Therefore, there are still some uncertainties with regard to synoptic-scale features as predictors of TCG. The confusion partly arises from a lack of understanding of the circumstances under which the combination of synoptic-scale features optimizes TCG. Moreover, use of these synoptic-scale features as TCG predictors in the Atlantic basin can be problematic because of the differences in basin size and landmass-ocean distribution. While the monsoon trough is the breeding region of most TCs in the western North Pacific basin, there is apparently no monsoon trough region in the western Atlantic [4].

In an analysis of Tropical Storm Arlene (2005), Yoo et al. [5, 6] found that low-level vortex dynamics advected temporary low-level westerly winds from the eastern North Pacific into the western Atlantic which, when combined with orographically enhanced low-level southeasterly winds from Central America, promoted TCG in the western Atlantic basin. Yoo [7] also suggested a potential influence of low-level wind enhancement in North America on several cases of western Atlantic TCG. Yoo [7] noted that when strong positive potential vorticity (PV) anomalies in the form of mid-latitude troughs occur with strong low-level convection over a vast region in the middle-to-high latitudes of North America, occasionally an alley of a low-level wind surge develops from the mid-latitude trough southward toward the western Atlantic, enhancing the large-scale low-level vortex of the developing storm. Yoo [7] also suggested that the enhancement of this low-level wind surge alley was a harbinger of the intensification of Hurricane Cindy and Hurricane Dennis of 2005. To better understand the relationship between large-scale geophysical features and TCG mechanisms over the western Atlantic, more case studies of TCG in the western Atlantic are warranted to analyze the characteristics of interactions of such features leading to TCG.

Hurricane Wilma (October 15–25, 2005) was the most intense hurricane recorded in the Atlantic basin, with a minimum central pressure of 882 hPa and an estimated peak sustained wind speed of 160 kt. Wilma caused 23 deaths and US$20.6 billion damage to the US alone [8]. Despite its record-breaking anomaly features, with the exception of a cursory analysis in the TC report from the National Hurricane Center (NHC), no study has been conducted to understand the relative importance of the atmospheric features discussed above in Wilma's TCG. This chapter reviews the large-scale atmospheric conditions from the early development stage of Wilma. The focus of the study is on the relative and collective roles of high-latitude PV anomaly and low-level wind surges of various origins.

2. Data and methods

The "best track" data from NHC are used as guidelines of the track and intensity changes of Hurricane Wilma. Large-scale sea surface temperature (SST) patterns are described using the National Oceanic and Atmospheric Administration (NOAA) optimum interpolation (OI) ¼ degree daily SST V2 data, which include in situ SST measurements from ships and buoys, satellite observations from Advanced Microwave Scanning Radiometer-Earth Observing System (AMSR-E) sensor on National Aeronautics and Space Administration (NASA's) Aqua satellite and NOAA 17/NOAA 18 Advanced Very High Resolution Radiometer (AVHRR), and National Centers for Environmental Prediction (NCEP) sea ice data [9].

The NCEP/National Center for Atmospheric Research (NCAR) reanalysis (NNR) data [10] are used to show the large-scale wind pattern during the development of Hurricane Wilma until it reached its peak maximum intensity over the Caribbean Sea by 1200 UTC October 19. The period of decreased intensity after moving from the Yucatan Peninsula to Florida will not be described in this study because the large-scale conditions that created the early TCG stage of Wilma are the main interest. The NNR dataset includes pressure-level variables in 17 vertical layers on a global 2.5° × 2.5° grid. Hennon and Hobgood [11] noted that NNR data are superior to the European Centre for Medium Range Weather Forecasts (ECMWF) reanalysis dataset in the tropics. Daily and long-term mean interpolated outgoing longwave radiation (OLR) data provided by the NOAA/OAR/ESRL PSD, Boulder, Colorado, USA, from their website at http://www.esrl.noaa.gov/psd/ are used to produce the daily OLR anomalies to represent strong surface convection [12].

The Weather Research and Forecasting (WRF) model is executed for Hurricane Wilma with a global model domain setting to reproduce hourly hemispheric-to-meso-scale circulations during its TCG and intensification period with improved spatial and temporal resolution over general circulation models (GCMs). The global WRF model domain was set with (x, y) dimensions of 721 × 361 (d01) (**Figure 1**). The global domain has a grid size of 55.58874 km in the x and y directions. The model microphysical schemes are configured following the NCAR Advanced Hurricane WRF (AHW) microphysics guidelines, including (i) Lin et al. cloud microphysics scheme [13]; (ii) the Rapid Radiative Transfer Model (RRTM) scheme for longwave radiation [14]; (iii) Dudhia scheme for shortwave radiation [15]; (iv) the Yonsei

University planetary boundary layer (PBL) parameterization; (v) the Monin-Obukhov scheme for the surface layer option; (vi) the thermal diffusion scheme for the land surface physics; and (vii) Kain-Fritsch (new Eta) scheme for the cumulus parameterization [16]. The 38 sigma (σ) level set of [17] was applied. The model "top" is defined at 50 hPa. The model run was set to update daily SST every 6 hours into the model integration. The daily "real-time global" (RTG) SST data were interpolated sequentially to produce 6 hourly input data for the WRF run.

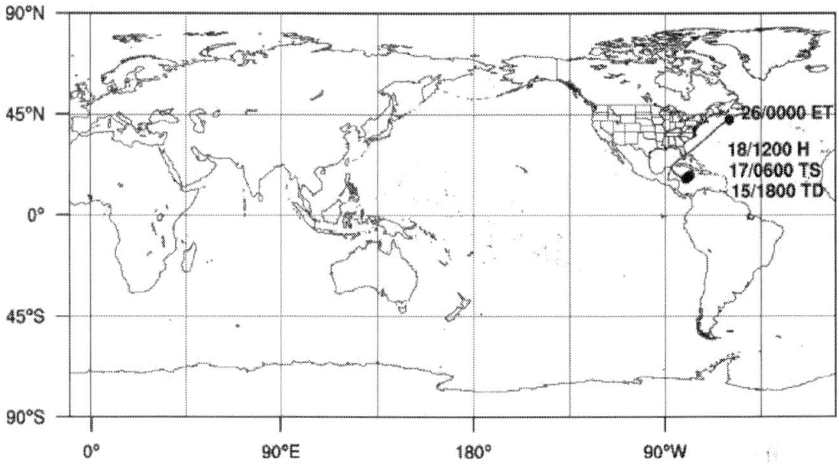

Figure 1. The model domain of global WRF for Hurricane Wilma in 2005. NHC best tracks were plotted with major intensity changes annotated.

In the global WRF model simulation, two different runtimes were executed. The first set of simulations runs from 0000 UTC October 7 to 0000 UTC October 21, to include the anomalous circulation that Hurricane Vince introduced in the North Atlantic about 1 week prior to Wilma's TCG—a duration of 336 hours. The second set of the simulations runs from 0000 UTC October 14 to 0000 21 October UTC, which only includes the pregenesis condition, TCG, and the development of Wilma—a duration of 168 hours. Both simulations end on 0000 UTC October 21, though Wilma continued to maintain hurricane intensity until 1800 UTC October 25. The latter half of the first model simulation result is compared to the result of the second run to evaluate the accuracy of the global WRF model.

To run the WRF model, 6 hourly NCEP Global Forecast System (GFS) final (FNL) operational global analysis data (1° × 1°) and daily RTG SST data are used for three-dimensional input data and for SST update (available from National Weather Service at ftp://ftp.polar.ncep.noaa.gov/pub/history/sst/), respectively. The model run is set to update SST every 6 hours into the model integration. The FNL data for this study are obtained from the Research Data Archive (RDA), which is maintained by the Computational and Information Systems Laboratory (CISL) at NCAR. The original data are available from the RDA (http://dss.ucar.edu) in dataset number ds083.2.

http://dx.doi.org/10.5772/64061

3. Hurricane Wilma in 2005 overview

3.1. Large-scale low-level dynamics

After Tropical Storm Tammy (October 5–6, 2005) diminished over Florida, a large-scale, low-level wind surge (>10 m s^{-1}) developed off the Atlantic coast of North America (**Figure 2a**). This surge was associated with a mid-latitude Rossby wave trough over the northern North

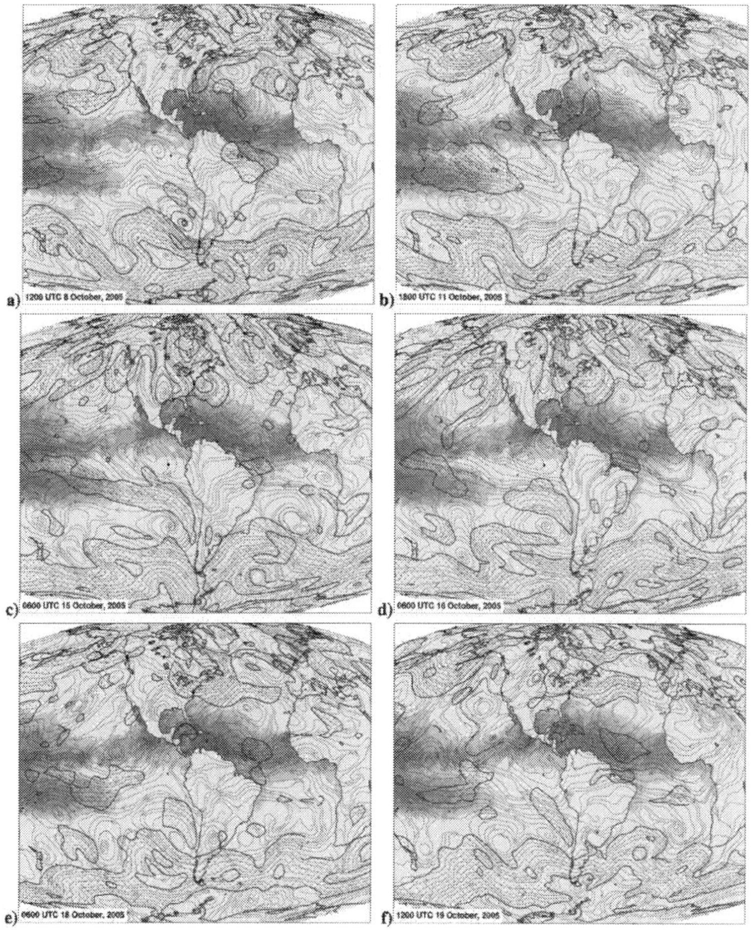

Figure 2. Wind stream analysis (850 hPa) of NCEP/NCAR reanalysis data and NOAA optimum interpolation (OI) ¼ degree daily sea surface temperature (SST) V2 data superimposed by low-level wind surge at (a) 1200 UTC October 8, (b) 1800 UTC October 11, (c) 0600 UTC October 15, (d) 0600 UTC October 16, (e) 0600 UTC October 18, and (f) 1200 UTC October 19, 2005. Areas of low-level wind surges exceeding 10 m s^{-1} are enclosed by dots within the thick solid contour lines, for which the interval is 15 m s^{-1}. SSTs exceeding 26°C are contoured with dark shades at a 1°C interval.

Atlantic and a subtropical anticyclone to the east of North America (also known as the Bermuda-Azores high). The mid-latitude trough seems to have assisted the formation of the subtropical high to the east of North America and a frontal low in the eastern North Atlantic, the latter of which later became Hurricane Vince (October 8–11, 2005) off the coast of western Africa (**Figure 2a**). Interactions between the subtropical high and the frontal low (Hurricane Vince) to its east deformed the shape of the subtropical high in its southern flank in the North Atlantic.

The abnormally strong frontal low also contributed to weakening of tropical easterly winds from western Africa. Due to the weakened easterly winds over the Caribbean Sea and the Gulf of Mexico during this period, the warm SST could avoid heat loss due to advection, accumulating more thermal energy in the low-level atmosphere and at the sea surface over the vast region. Meanwhile, vigorous southeasterly winds emanating from the South Atlantic subtropical high produced strong low-level wind surges over a vast area northeast of Brazil (**Figure 2a**).

Regarding the development of Hurricane Wilma, low-level westerly winds from the eastern North Pacific should not be disregarded. The evolution of synoptic-scale low-level vortices in the eastern North Pacific is driven by the interplay between subtropical high pressure systems in the eastern South Pacific and in the eastern North Pacific, at least in the case of Arlene four months earlier [6]. Since early in October, synoptic-scale low-level conditions in the Pacific had supported the development of a low-level anticyclonic vortex in the eastern North Pacific. In particular, anomalously strong southeasterly winds from the southeastern Pacific subtropical high advected momentum effectively to the anticyclone in the eastern North Pacific, culminating in low-level westerly winds in the region (**Figure 2b**). Although the intensities of the low-level wind surges changed slightly, the general setting of the large-scale low-level vortices distribution around Central America was maintained for an extended period of time during early October.

Particularly, the western flank of the southeastern protrusion of the elongated North Atlantic subtropical anticyclone, which was northeast of the Caribbean Sea or Puerto Rico, became a seed zone for the development of an unusually intense 850 hPa cyclonic vortex at the subsynoptic scale by October 8 (**Figure 2a** and **b**). This vortex was able to develop as the easterly wind in the central tropical Atlantic weakened, allowing the cross-equatorial southeasterly wind east of the Caribbean Sea to approach 25°N (**Figure 2a**). By October 9, the abnormally large-scale low-level cyclone with its center hovering over the eastern Caribbean maintained its extended shape, from the southwestern Caribbean to the northeast.

Meanwhile, the westerly winds from the eastern North Pacific were sufficiently strong to traverse the western Caribbean around October 11, and they converged with the southeasterly and easterly winds in the eastern Caribbean (**Figure 2b**). This convergence reinforced the preexisting subsynoptic-scale low-level cyclonic flow over the Caribbean Sea, which continued on 0000 UTC October 13 despite weakened westerly winds upon the disappearance of the eastern North Pacific low-level circulation.

Beginning around 0000 UTC October 14, a meso-scale high and a mid-latitude trough were developing over Texas and inland central Canada, respectively. This anticyclone over Texas grew quickly while the subsynoptic-scale low-level cyclonic flow over the Caribbean Sea had split into two cyclonic circulations over the Caribbean and adjacent to the US Atlantic coast—one on the northern and the other on the southern edge of the cyclonic zone (**Figure 2c**). The northern cyclonic circulation near the US Atlantic coast became an extratropical cyclone, and the southern cyclone over the Caribbean Sea strengthened near Jamaica by October 14, becoming a tropical depression by 1800 UTC October 15. Pasch et al. [8] suggested that tropical waves "traversing the Caribbean" might have been associated with the formation of the tropical depression. However, no apparent tropical wave "traversing" the Caribbean during that time might have affected the TCG. Instead, the 850 hPa streamline analysis shows clearly that the large-scale low-level wind was traversing the Caribbean from north to south (**Figure 2c**). By 1200 UTC October 15, the mid-latitude trough merged with the extratropical cyclone, strengthening the extratropical cyclone further (**Figure 2d**). The enhanced extratropical cyclone impeded low-level easterly winds from the central tropical Atlantic from entering into the Caribbean Sea by deflecting the easterlies in the central tropical Atlantic northward to the east of the Caribbean Sea. This circulation allowed for the sustenance of the low-level circulation in the Caribbean Sea without significant interference by the normally zonally propagating tropical waves, allowing the warm SST to accumulate more thermal energy at the sea surface and adjacent low-level atmosphere over the Caribbean Sea (**Figure 2d**).

Meanwhile, the tropical depression over the Caribbean Sea maintained its subsynoptic-scale vortex between the northerly winds associated with the trough in the northeastern US and the southeasterly winds from the South Atlantic (**Figure 2d**). Neither of the two 850 hPa circulations were strong enough to support a wind surge around the intensifying tropical depression, but these two moderate drive trains caused the "weak and ill-defined steering" [8] of the storm for the first few days of the storm that was to become Wilma. In this environment of only weak background and adjacent circulations, the depression slowly strengthened over October 16 and became a tropical storm at 0600 UTC October 17.

Over October 17–18, the North Atlantic subtropical high strengthened and began to produce more vigorous low-level easterly winds in the central tropical Atlantic toward the Caribbean Sea (**Figure 2e**). At the same time, the mid-latitude trough continued to advect the momentum of the low-level atmosphere to the anticyclone centered over Texas. This anticyclone continued to produce 850 hPa northerly winds from the southeastern US toward the Yucatan peninsula, to the west of Wilma in the Caribbean Sea, thereby enhancing cyclonic vorticity of Wilma (**Figure 2e**). During that time, Wilma drifted toward the west-northwest and strengthened into a hurricane by 1200 UTC October 18. An explosive deepening occurred during the night of the 18th/19th, and Wilma's maximum sustained wind speed had increased to near 150 kt (Category 5 on the Saffir-Simpson Hurricane Scale) by 0600 UTC October 19 [8]. By 1200 UTC October 19, the peak sustained wind speed of 160 kt was recorded for Wilma with the estimated minimum central pressure of 882 hPa—the lowest pressure recorded for a hurricane in the Atlantic basin [8]. While Wilma underwent this unprecedented rapid intensification, low-level wind surges associated with the North Atlantic high were active over the northeastern US and in the central tropical Atlantic (**Figure 2f**). Hurricane Wilma was located in the middle of the steering flow cornered by the terrain in Central America.

3.2. Sea surface temperature conditions

As shown in **Figure 3**, SSTs in the entire western Atlantic basin and central tropical Atlantic exceeded the 26°C climatological threshold for TC development (e.g. [18]) during the lifespan of Wilma, with western Atlantic SSTs peaking at over 30°C. In contrast, SSTs in the eastern North Pacific basin were barely over 26°C (**Figure 3**). The general SST condition over the western Atlantic was warmer than the average during 1971–2000, while the southern Caribbean and central tropical Atlantic had even stronger positive anomalies during the period (**Figure 4**). It is notable that SSTs were below average over the majority of the eastern North

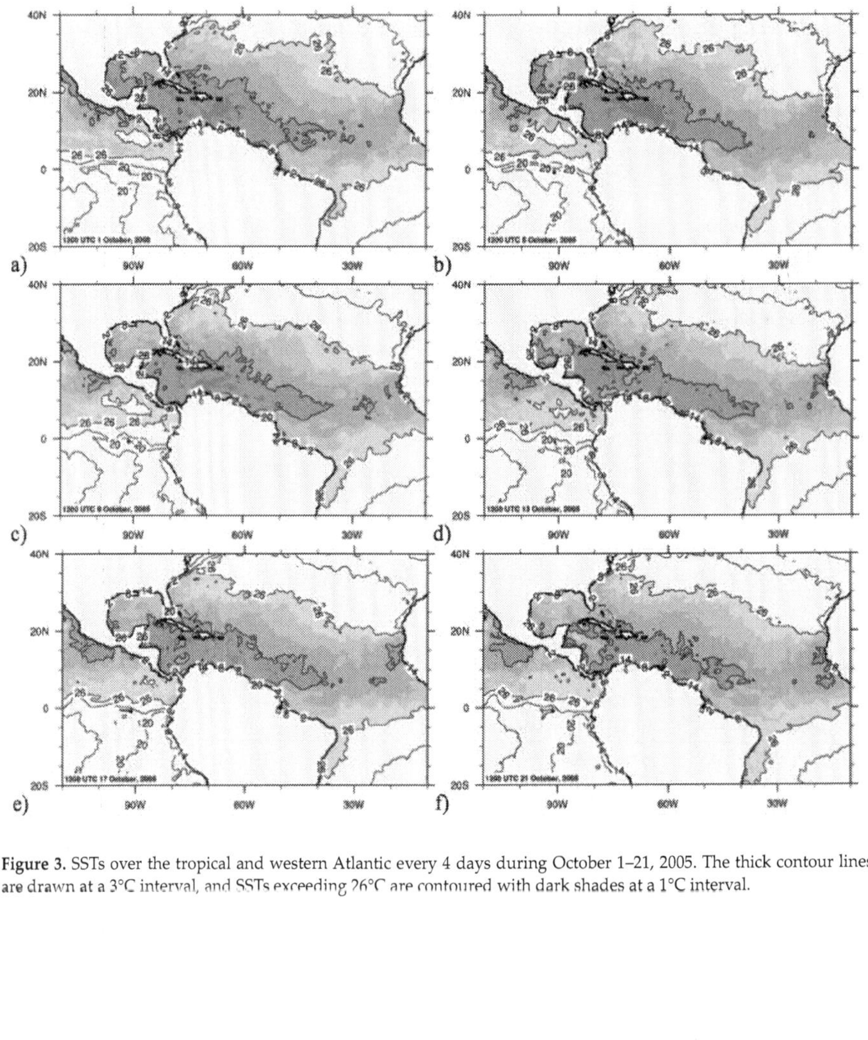

Figure 3. SSTs over the tropical and western Atlantic every 4 days during October 1–21, 2005. The thick contour lines are drawn at a 3°C interval, and SSTs exceeding 26°C are contoured with dark shades at a 1°C interval.

http://dx.doi.org/10.5772/64061

Pacific during the same period except for some positive anomaly hot spots along the equator (**Figure 4c–f**).

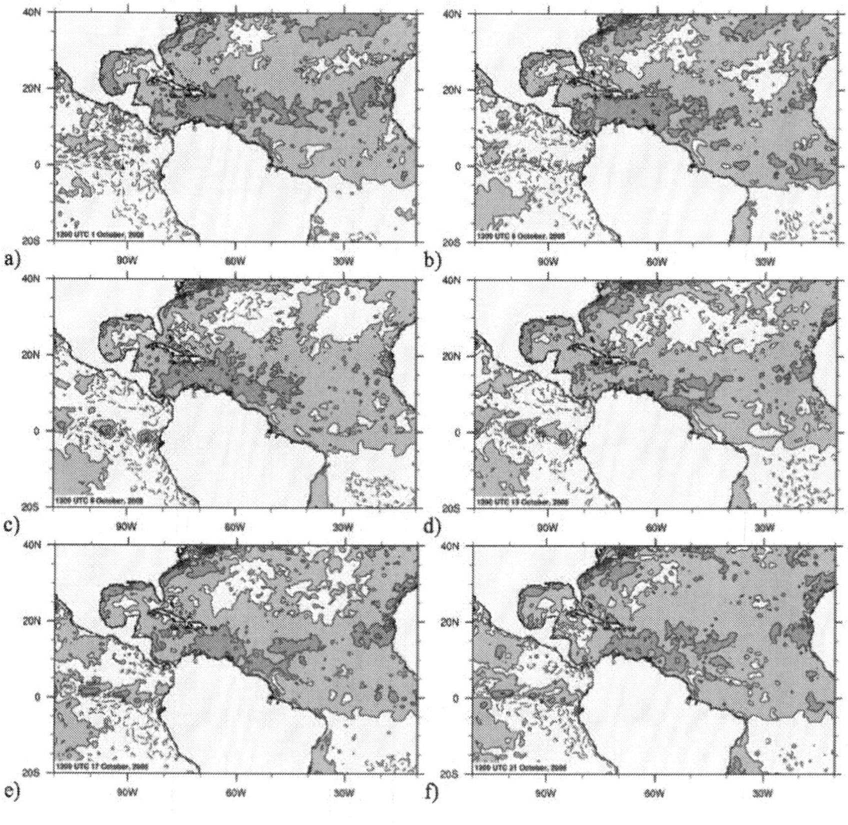

Figure 4. SST anomalies over the tropical and western Atlantic every 4 days during October 1–21, 2005. Contour lines are drawn at a 1°C interval. Positive anomalies are contoured with dark shades at a 1°C interval, and negative anomalies are drawn with dashed line contours.

SST changes over 3 and 6 days are depicted in **Figures 5** and **6**, respectively. Generally, SSTs over the Atlantic remained constant or increased slightly. Interestingly, 6-day SST change maps clearly show significant SST decreases in the Atlantic during October. SST decrease during October 1–7 (**Figure 6a**) over a broad area in the northern North Atlantic seems to be related to the passage of a mid-latitude trough that was associated with Tropical Storm Tammy (October 5–6; see **Figure 2a**). The decreasing SST in the Caribbean during October 7–13 (**Figure 6b**) represents the evaporation from the ocean, while the tropical depression (Wilma) was developing. The SST change map for October 13–19, 2005 (**Figure 6c**) suggests sea surface energy consumption by Hurricane Wilma during its early explosive intensification in the northwestern Caribbean Sea. Finally, the SST decrease in the central tropical Atlantic (**Figure 6d**) provides evidence of the vigorous low-level southeasterly and easterly wind surges during October 19–25, 2005 (see **Figure 2f**).

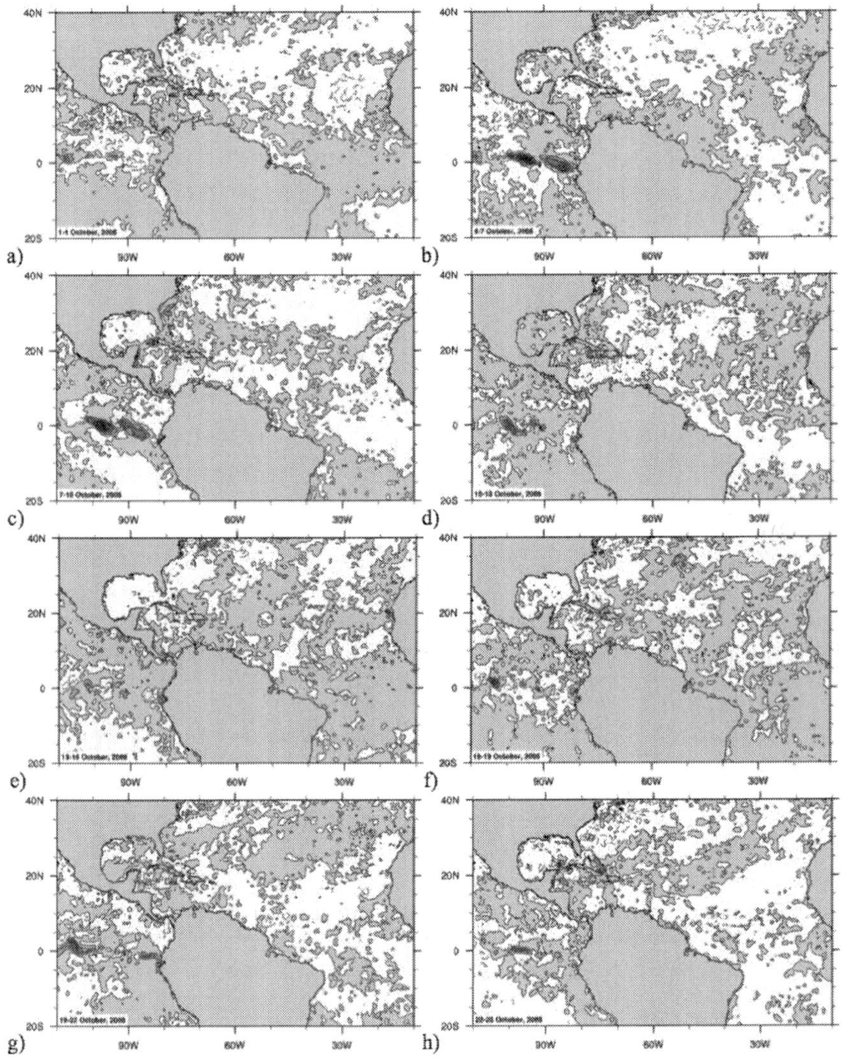

Figure 5. SSTs change over 3-day intervals over the tropical and western Atlantic during October 1–25, 2005. Contour lines are drawn at a 1°C interval. SST increases are contoured with dark shades at a 1°C interval, and SST decreases are drawn with dashed line contours.

http://dx.doi.org/10.5772/64061

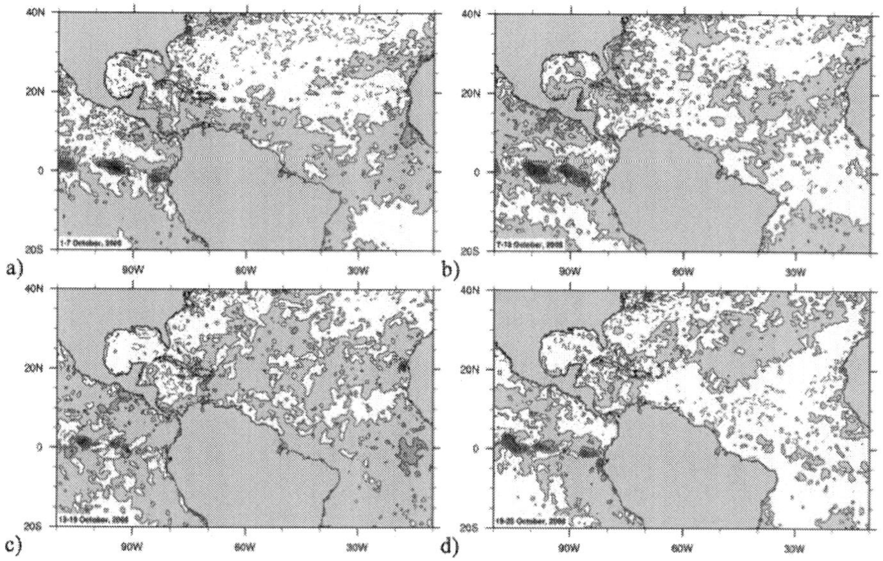

Figure 6. As in **Figure 5**, except for 6-day changes.

3.3. Potential vorticity in the upper-level atmosphere and outgoing longwave radiation

In general, mid-latitude troughs are associated with lower-level instability downwind of the trough as they provide momentum from the high latitudes and incorporate moisture from the lower latitudes. In the mid-latitudes, upper-tropospheric troughs in the Rossby waves have positive PV by nature, and positive PV anomaly regions correspond well to the locations of the 850 hPa wind surges which seem to be associated with convection as shown by OLR anomalies (**Figures 7** and **8**). However, those relationships do not apply to the lower-latitude region. Generally, upper-level PV is significantly weaker in the lower latitudes than in the mid-latitudes. Nevertheless, strong convection can occur there even with mild wind speeds, mainly because SST conditions tend to be much more favorable for strong convection in the lower latitudes than in the mid-latitudes (**Figure 3**). **Figure 8** clearly shows the cyclonic circulation development of Wilma with strong OLR anomalies over the Caribbean.

The mid-latitude trough that merged with an extratropical cyclone around the TCG period of Hurricane Wilma on October 15 (**Figure 2c** and **d**) was accompanied by the anomalously latitudinally stretched Rossby waves in the Northern Hemisphere during the period. This peculiar formation of Rossby waves resulted in a rather unusual synoptic-scale low-level wind flow with a sharp meridional circulation pattern as well as an abnormal 200 hPa PV distribution over North America (**Figure 7**). The unusual deformation of Rossby waves is likely to have increased atmospheric momentum and turbulence in the tropical latitudes by advecting vorticity during Wilma's TCG.

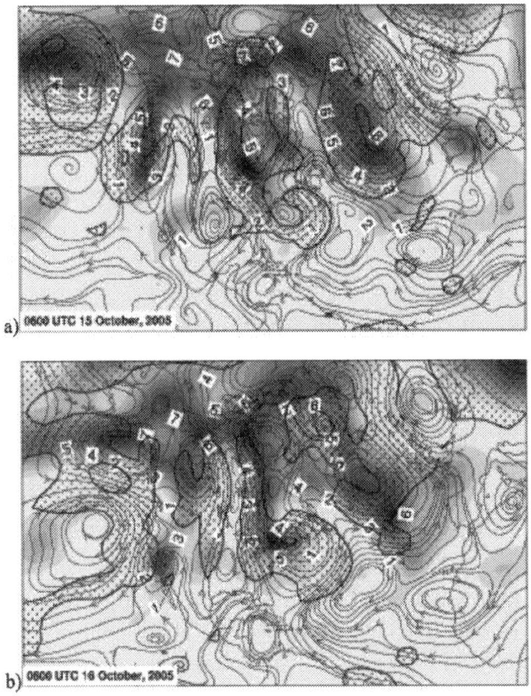

Figure 7. Abnormal distribution of 200 hPa potential vorticity (PV, shaded, 10^{-6} K m^2 kg^{-1} s^{-1}) during the early TCG period of Wilma (October 15–16, 2005). Streamline analysis (850 hPa) is superimposed on the low-level wind surge. Areas of 850 hPa wind surges exceeding 10 m s^{-1} are enclosed by dots within the thick solid contour lines, for which the interval is 15 m s^{-1}.

During Wilma's explosive deepening (from 1200 UTC October 18 to 1200 UTC October 19), strongly positive PV and a 850 hPa trough were over the northeastern coast of the US accompanied by extremely large low-level wind surge regions from the central US to the North Atlantic (**Figure 8d** and **e**). It is interesting to note how dynamic the low-level Northern Hemisphere atmosphere was overall, as evidenced by the low-level wind surges during the TCG and intensification period (**Figure 8a–e**). Bracken and Bosart [19] suggested that vorticity advection between a subtropical anticyclone and a developing storm can play an important role in accelerating TCG. Thus, the distribution of cyclones and anticyclones affects the evolution of each circulation cell through their interplay, advecting vorticities and momentum. Therefore, the synoptic-scale anticyclone over Texas and the subtropical high in the North Atlantic should have also assisted in the spin-up of Wilma over the Caribbean Sea. The clockwise low-level circulation of the Texas anticyclone to the northeast of Wilma made contact over the Gulf of Mexico, while the North Atlantic subtropical high was interacting with Wilma, mainly in the form of easterly winds. Both anticyclones seem to have contributed to Wilma's intensification at the 850 hPa level by advecting angular momentum to the outer radii of Wilma.

http://dx.doi.org/10.5772/64061

The tropical easterly winds associated with the North Atlantic high contributed to Wilma's intensification by supplying a substantial amount of enthalpy from the central tropical Atlantic during October 19 (see **Figure 6**).

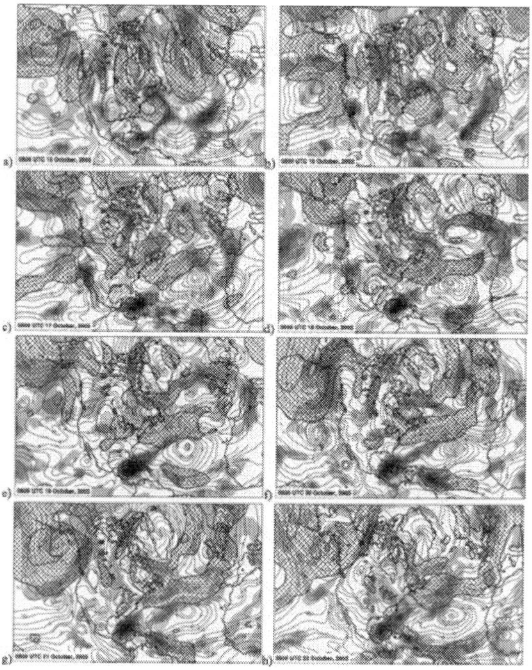

Figure 8. Evolution of the distribution of negative OLR anomalies (W m^{-2}, shaded) during October 15–22, 2005. Streamline analysis (850 hPa) is superimposed on the low-level wind surge. The contour interval for OLR anomalies is 30 W m^{-2}. Areas of 850 hPa wind surges exceeding 10 m s^{-1} are enclosed by dots within the thick solid contour lines, for which the interval is 15 m s^{-1}.

However, as the regional anticyclone over Texas weakened, Wilma's intensity decreased on October 20 by 30 kt (still leaving her as a Category 4 hurricane) from 160 kt at 1200 UTC October 19, although at that time the tropical easterly winds from the central tropical Atlantic strengthened. These events suggest that the large-scale circulation pattern in the immediate TC environment plays an important role in TC intensity change [20]. The multidirectional sources of angular momentum advection as described here likely provided for more efficient intensification than if the angular momentum inflow had been from a sole source, such as low-level easterlies. Meanwhile, the negative OLR anomaly maps after Wilma's TCG show that Hurricane Wilma's explosive deepening (from 1200 UTC October 18 to 1200 UTC October 19) was favored by the persistent low-level inflow from the large convective region in the Caribbean Sea to the central tropical Atlantic, which is consistent with the SST change over the period (**Figures 5** and **6**).

4. WRF model simulation

Figure 9 shows wind conditions at 1500 m (about 850 hPa) at the time of model integration. Compared to **Figure 2**, the global WRF reproduced large-scale atmospheric circulations very closely and effectively. The color of the wind arrows represents magnitudes of wind speed; wind speed increases as the color changes from red to yellow to green to blue (0–46 m s^{-1}). Arrows in green represent winds exceeding 10 m s^{-1}. The yellow-to-purple scale in the background represents the 1500 m high water vapor in the range of 12–20 g kg^{-1}. The deformation of the low-level circulation by Hurricane Vince (October 8–11; **Figure 9a**) and the subsequent development of the anomalous low-level cyclone over and northeast of the Caribbean Sea by the westerly winds from eastern North Pacific were simulated very realistically (**Figure 9b**).

Figure 9. Samples of global WRF model simulation outputs: the anomalous circulation development of the subtropical high in the North Atlantic due to Hurricane Vince (October 8–11) (a) and the anomalous low-level cyclone over, and northeast of, the Caribbean by the westerly winds from eastern North Pacific (b). Wind conditions at 1500 m (about 850 hPa) at the time of model integration are presented with arrows. The color of the wind arrows represents wind speed: wind speed increases as the color changes from red to blue (0–46 m s^{-1}). The yellow to purple in the background represents the water vapor in the range of 12–20 g kg^{-1}.

http://dx.doi.org/10.5772/64061

However, the global WRF model simulation deteriorated by the seventh day after initialization (**Figure 10**), preventing the model from replicating the TCG of Wilma and its subsequent development into a hurricane over the western North Atlantic (not shown). From **Figure 10**, it seems that the major errors occurred in the high latitudes and mid-latitude systems that are directly affected by the high-latitude circulations. In fact, during the prolonged global WRF simulation, the major failure occurred in reproducing the interactions between the mid-latitude trough over the northeastern US and the subtropical low to the east of the US. As a result, the erroneous forecast in intensity and location of the mid-latitude trough over the northeastern US that was positioned to the northeast of the disturbance in the Caribbean predicted that the subtropical high would be restored in the North Atlantic, and vigorous easterly winds would resume over Central America (**Figure 11**) by October 16–20. This zonally enhanced low-level wind condition is opposite to the meridionally enhanced, low-level condition in the actual case of Wilma. In fact, the restored easterly winds over the Caribbean Sea impeded the development of a storm in the region.

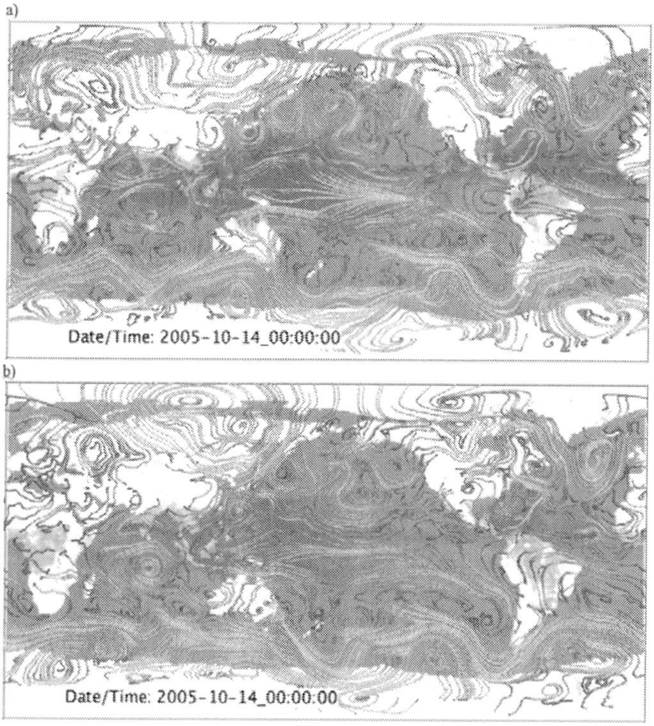

Figure 10. Comparison of the large-scale low-level circulations between the global WRF model results after a 7-day continuous simulation (a) and at the model initialization (b) both at 0000 UTC October 14 using FNL and RTG SST datasets.

Figure 11. A failed forecast of Wilma's TCG.

By contrast, the second simulation that was initialized at 0000 UTC October 14 successfully simulated the TCG of Wilma and its subsequent development (**Figure 12**). The WRF global model reproduced every major vortex and circulation at 850 hPa level not only over the North Atlantic but also over the neighboring basins, including the eastern Pacific, South Atlantic, and subpolar regions. But the forecasted track of Wilma shifted to the east from the actual best track when Wilma was in its hurricane stage. This error seems to be attributed to the use of a relatively large grid size for the model simulation (about 56 km) to represent the meso-scale features accurately, while the inner-core dynamics of the storm actually might have played a more important role in steering its path and determining its intensity.

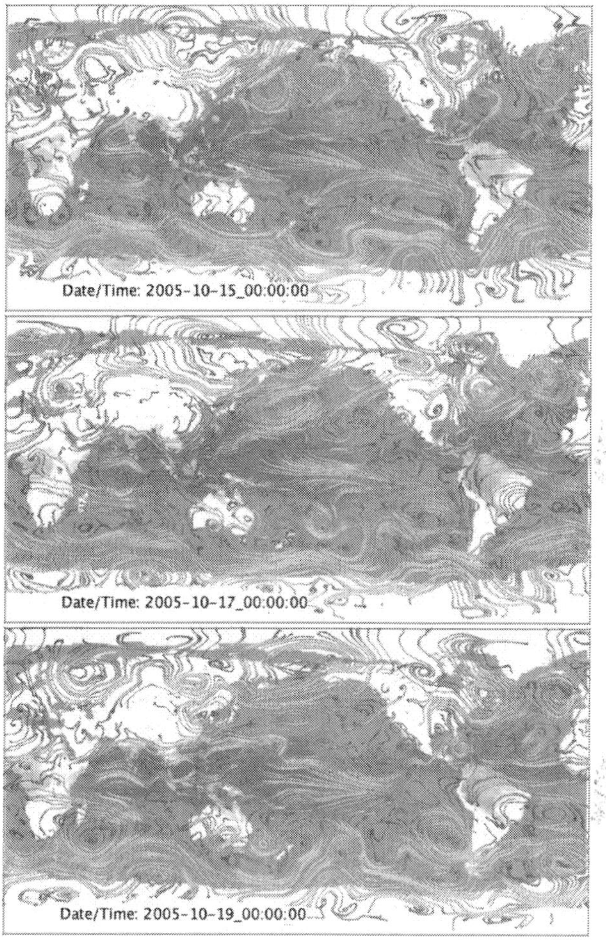

Figure 12. A successful forecast of Wilma's TCG.

The successful forecast of the merger of the subtropical cyclone with the mid-latitude trough off the east coast of the US around October 16 seems to be the key point that led the subsequent successful forecasts in the unique low-level, large-scale circulation development in the case of Wilma (see **Figures 3–6** and **12**). This result suggests that the role of mid-latitude systems in TC activity is not negligible; a result that is incongruent with the current TCG forecasting emphasis solely on tropical atmospheric conditions. The comparison of the unsuccessful and successful forecasts of the TCG and development of Wilma suggests that every major vortex and circulation component at least in the immediately neighboring storm development area is important for TCG progression.

Although the calculation errors in the WRF global model simulation grew significantly after seven days of model integration, it seems that the global WRF can be used for the purpose of operational short-range TCG forecasting. This result is encouraging, considering the fact that the global WRF model was initialized 42 hours before the simulated TCG and subsequent development of a TC, and yet the forecast will be fairly accurate in one continuous simulation for the 7-day period. It should be also noted that the 7-day global WRF model simulation required less than six hours in a Linux cluster computer with 96 cores.

5. Summary and conclusion

Planetary-scale atmospheric and oceanic conditions of the western Atlantic basin were analyzed to understand the unique TCG and intensification mechanism of Hurricane Wilma in 2005, using NCEP/NCAR reanalysis data, NOAA optimum interpolation (OI) ¼ degree daily SST V2 data, NOAA/OAR/ESRL PSD interpolated OLR data, and global WRF model simulation. An anomalous development of the 850 hPa circulation pattern in the North Atlantic was triggered by Hurricane Vince (October 8–11, 2005) in the eastern North Atlantic. Circulation around the southeastern fringe of the North Atlantic subtropical anticyclone during the period had been interrupted by the presence of Vince, causing a perturbation in the downstream flow around the entire southern edge of the North Atlantic subtropical high. On the southwestern flank of the subtropical high, the perturbation contributed to the development of a large-scale 850 hPa vortex, which would eventually allow for Wilma's TCG in the eastern Caribbean Sea. Due to the change in the low-level circulation by the deformed subtropical anticyclone, weakened low-level easterly winds allowed southeasterly winds from the Southern Hemisphere and westerly winds from eastern North Pacific to become relatively important, generating an anomalously prominent low-level cyclone over the western Atlantic about a week before TCG.

The anomalously large low-level cyclone over the western Atlantic matured over the warm ocean before it was separated into two cyclones in a north-south alignment (a northern cyclone and a southern cyclone). The separation was caused by the advance of northerly winds from a mid-latitude trough over central Canada one day before TCG. By 1200 UTC October 15, the high-latitude trough merged with the northern cyclone, resulting in a strengthened northern subtropical low. The enhanced subtropical low eventually played a role in sustaining the low-level circulation in the Caribbean Sea by preventing a significant interference from the zonally propagating tropical waves (**Figure 2c** and **d**). The southern cyclone became more concentrated in the Caribbean Sea, near Jamaica by October 14, growing into a tropical depression by 1800 UTC October 15.

The unusual but persistent meridionally oriented circulation conditions allowed the tropical depression over the Caribbean Sea to strengthen slowly between the northerly winds associated with the trough in the northeastern US and the southeasterly winds from the South Atlantic (**Figure 2c** and **d**). Wilma became a tropical storm at 0600 UTC October 17. Over October 17–18, as the North Atlantic subtropical high strengthened to produce more vigorous

http://dx.doi.org/10.5772/64061

low-level easterly winds in the central tropical Atlantic toward the Caribbean Sea (**Figure 2e**), Wilma drifted toward the west-northwest and strengthened into a hurricane at 1200 UTC October 18.

The unprecedented rapid intensification of Hurricane Wilma during the next night took place under anomalously warm SST conditions when Wilma was trapped between the northerly winds from the mid-latitude trough and synoptic-scale southeasterly winds. During Wilma's explosive deepening, a synoptic-scale anticyclone over Texas and the North Atlantic subtropical high seems to have caused Wilma to intensify between them by advecting angular momentum to the outer radii of Wilma.

The global WRF reproduced planetary-scale atmospheric circulations at 1500 m (about 850 hPa) very closely and effectively, including the deformation of the low-level circulation by Hurricane Vince (October 8–11) and the subsequent development of the anomalous low-level cyclone over and northeast of the Caribbean. However, the error growth after seven days from the model initialization changed the steering circulations, resulting in a failed forecast of Wilma's TCG and its subsequent development into a hurricane over the western North Atlantic. It seems that the major error resulted from misrepresentation of the interactions between mid-/high-latitude systems and tropical circulations.

In contrast, the second simulation that was initialized at 0000 UTC October 14 successfully simulated Wilma's TCG and subsequent development (**Figure 12**). The WRF global model reproduced every major vortex and circulation at the 850 hPa level, not only over the North Atlantic but also at least over the neighboring ocean basins. With the successful simulation of the merger of the subtropical cyclone with the mid-latitude trough off the east coast of the US around October 16, the subsequent forecast of the global WRF model was maintained successfully, reproducing the unique low-level, large-scale circulation development in the case of Wilma (see **Figures 2c–f** and **12**). The result of the global WRF model suggests that the role of mid-latitude systems in TC activity is more important than previously considered, and that every large-scale vortex and circulation component at least in the immediately neighboring region of the storm developing area is important for TCG forecasting. More in-depth analyses are warranted to better understand the unusual development of Hurricane Wilma.

Author details

Jinwoong Yoo[1*], Robert V. Rohli[2] and Jennifer Collins[3]

*Address all correspondence to: jinwoong.yoo@gmail.com

1 The Purdue Climate Change Research Center, Purdue University, West Lafayette, IN, USA

2 Department of Geography and Anthropology, Louisiana State University, Baton Rouge, LA, USA

3 School of Geosciences, University of South Florida, Tampa, FL, USA

References

[1] McBride, J.L., and R. Zehr, 1981: Observational analysis of tropical cyclone formation. Part II: Comparison of non-developing versus developing systems. J. Atmos. Sci., 38, 1132–1151.

[2] Briegel, L.M. and W.M. Frank, 1997: Large-scale influences on tropical cyclogenesis in the western North Pacific. Mon. Wea. Rev., 125, 1397–1413.

[3] Gray, W.M., 1998: The formation of tropical cyclones. Meteorol. Atmos. Phys., 67, 37–69.

[4] Chan, J.C. and R.H. Kwok, 1999: Tropical cyclone genesis in a global numerical weather prediction model. Mon. Wea. Rev., 127, 611–624.

[5] Yoo, J., J.M. Collins, and R.V. Rohli, 2014: Tropical cyclogenesis in the Intra-Americas Sea: Hurricane Cindy (2005). Prof. Geogr., 66, 511–524.

[6] Yoo, J., J.M. Collins, and R.V. Rohli, 2015: An investigation of the tropical cyclogenesis of Arlene (2005) using the ERA-Interim reanalysis and the WRF model simulation. Prof. Geogr., 67, 396–411.

[7] Yoo, J., 2011: Large-scale influences on tropical cyclogenesis for selected storms in the 2005 Atlantic hurricane season, Ph.D. Dissertation, Louisiana State University.

[8] Pasch, R.J., E.S. Blake, H.D. Cobb, and D.P. Roberts, 2006: Tropical Cyclone Report Hurricane Wilma. National Hurricane Center.

[9] Reynolds, R.W., T.M. Smith, C. Liu, D.B. Chelton, K.S. Casey, and M.G. Schlax, 2007: Daily high-resolution blended analyses for sea surface temperature. J. Climate, 20, 5473–5496.

[10] Kalnay, E., M. Kanamitsu, R. Kistler, W. Collins, D. Deaven, L. Gandin, M. Iredell, S. Saha, G. White, J. Woollen, Y. Zhu, M. Chelliah, W. Ebisuzaki, W. Higgins, J. Janowiak, K.C. Mo, C. Ropelewski, J. Wang, A. Leetmaa, R. Reynolds, R. Jenne, and D. Joseph, 1996: The NCEP/NCAR 40-year reanalysis project. B. Am. Meteorol. Soc., 77, 437–470.

[11] Hennon, C., and J.S. Hobgood, 2003: Forecasting tropical cyclogenesis over the Atlantic basin using large-scale data. Mon. Wea. Rev., 131, 2927–2940.

[12] Liebmann, B., and C. Smith, 1996: Description of a complete (interpolated) outgoing longwave radiation dataset. B. Am. Meteorol. Soc., 77, 1275–1277.

[13] Lin, Y.-L., R.D. Farley, and H.D. Orville, 1983: Bulk paramerterization of the snow field in a cloud model. J. Appl. Meteor., 22, 1065–1092.

[14] Mlawer, E.J., S.J. Taubman, P.D. Brown, M.J. Iacono, and S.A. Clough, 1997: Radiative transfer for inhomogeneous atmospheres: RRTM, a validated correlated-k model for the longwave. J. Geophys. Res., 102, 16663–16682.

[15] Dudhia, J., 1989: Numerical study of convection observed during the winter monsoon experiment using a meso-scale two-dimensional model. J. Atmos. Sci., 46, 3077–3107.

[16] Gilliland, E.K., and C.M. Rowe, 2007: A comparison of cumulus parameterization scheme in the WRF model. In: Proceedings of the 87th AMS Annual Meeting & 21st Conference on Hydrology, (P2.16). San Antonio, TX, USA.

[17] Kieu, C.Q., and D.-L. Zhang, 2008: Genesis of tropical storm Eugene (2005) from merging vortices associated with ITCZ breakdowns. Part I: Observational and modeling analyses. J. Atmos. Sci., 65, 3419–3439.

[18] Gray, W.M.,1968: Global view of the origin of tropical disturbances and storms. Mon. Wea. Rev., 96, 669–700.

[19] Bracken, W. E., and L. F. Bosart, 2000: The role of synoptic-scale flow during tropical cyclogenesis over the North Atlantic Ocean. Mon. Wea. Rev., 128, 353–376.

[20] Miller, B.I., 1958: On the maximum intensity of hurricanes. J. Meteorol., 15, 184–195.

Chapter 6

Mesoscale Convective Systems and Early Development of Tropical Cyclones

Kevin K. W. Cheung and Guoping Zhang

Additional information is available at the end of the chapter

http://dx.doi.org/10.5772/64185

Abstract

This chapter studies two Tropical cyclone (TC) cases, Typhoon Dan (1999) and Typhoon Ketsana (2003), and discusses their rates of formation and relationship with the mesoscale convective activities through examining the numerical simulations of the two cases. Many TCs generate from a single mesoscale convective System (MCS) or multiple MCSs; the physical processes under these two patterns are found to include dissipation of convection leading to new eruptions of deep convection located near the edge of the dissipating convection core, ingestion of nearby convection, merging of multiple MCSs into one MCS, and merging of deep convection within the MCS associated with the aggregation of vorticity in early development stage of TCs. How these activities lead to the formation of Typhoon Ketsana has been diagnosed. The diabatic heating associated with these convective activities also help to form the TC warm core. The relationship between the rate of TC formation and early development and convection energy consumption is discussed.

Keywords: tropical cyclone formation, mesoscale convective systems, stratiform and convective rain, diabatic heating, convective available potential energy

1. Introduction

Over the recent decades, researches for the process that generates a surface vortex have focused on the observation that TC formation is associated with mesoscale convective systems (MCSs) and their accompanying mesoscale convective vortices (MCVs). It was believed that the transition from MCS to a TC-like vortex required the generation of low-level cyclonic vorticity below the MCS, and researches for TC genesis mechanisms focus on what provided this sub-MCS low-level cyclonic vorticity. It is common to observe MCS or MCSs involved in the formation process of TCs in the western North Pacific (WNP). Over 63% of the studied TCs have multiple MCS convection appearing at the formation process, and 35% have the single MCS appearing at the beginning of the formation process. About 90% of cases with a shorter time period of formation process (within 6 h) have single MCS [1]. The physical processes under the single MCS and multiple MCS convection are found to include dissipation of convection leading to new eruptions of deep convection located near the edge of the dissipating convection core, ingestion of nearby convection, merging of multi-MCSs into single MCS, and merging of deep convection within the MCS associated with the aggregation of vorticity in fast formation process.

An MCS is organized as convective cloud clusters, and TC formation process is due to increasing organizition of MCS, which is caused by accumulation of mesoscale vorticity [2]. Convective activities resulted from thermodynamic and/or dynamic effects that affect the evolution of early development of TCs [3, 4]. They play critical roles in the formation process on TCs; furthermore, the local warming and diabatic heating associated with these convective activities also help to spin up the circulation and generate the warm core structure. How the pattern of MCS convection affects the rate of TC formation and early development is the focus of this chapter. In this chapter, two TC cases are studied and their synopses are first provided before their rates of development are discussed. The model simulations of the two cases are validated in terms of synoptic development and convective episodes. The convection types in the model are separated into convective and stratiform types with their respective vertical heating profiles. Then the heating associated with the MCSs and its effect on TC development are analyzed based on previous theories [5] and through Eliassen–Palm (EP) flux analysis. Then, the rates of development in the two TC cases are discussed based on the convective energy consumption point of view. In regard to this view, convection development and maintenance consume convective available potential energy (CAPE) in their local environment. It gradually makes the local environment less conducive for further convection development, which may affect the temporal evolution of TC formation when the surface vorticity is still below the threshold for tropical storm. However, it humidifies the middle and upper troposphere, and then gradually builds up the value of CAPE again in the local environment until new deep convection bursts up.

2. Typhoon cases and numerical simulations

The numerical experiments in this study include two case studies: Typhoon (TY) Ketsana (2003) with the development of multiple MCSs and Typhoon Dan (1999) with only one MCS involved during its formation. These experiments are then examined to study the difference in the formation time and energy consumption in the two cases. In the simulations for both typhoon cases, the same set of nested domain settings is applied, and the same satellite datasets have been assimilated via the Weather Research and Forecasting Data Assimilation (WRFDA) system.

In the WRF simulation of Typhoon Ketsana, the model's initial and lateral conditions are taken from the National Center for Environmental Prediction (NCEP) final (FNL) analysis data with the outermost lateral boundary conditions updated every 6 h, and the SST data are interpolated to update the sea-surface boundary conditions. The WRF model is initialized at 0000 UTC on October 16, 2003 and integrated for 108 h until 1200 UTC on October 20. Within this model integration period, the WRFDA is used to assimilate QuikSCAT oceanic winds available at 0600 UTC on October 16 and 0600 UTC on October 17, 2003, and SSM/I oceanic surface wind speed and total precipitable water (TPW) available at 1200 UTC on October 17, 2003 into the boundary conditions. These twice daily swaths of QuikSCAT and SSM/I data are extracted from the remote sensing system (RSS) data archive with 0.25° latitude/longitude resolution.

The WRF simulation of Typhoon Dan (1999) is initialized at 0000 UTC on October 1, 1999 and integrated for 120 h until 0000 UTC on October 6. Similar to the case of Typhoon Ketsana, the WRFDA is used to assimilate QuikSCAT oceanic winds available at 0600 UTC on October 1 and 0600 UTC on October 2, and SSM/I oceanic surface wind speed and TPW available at 1200 UTC on October 1 and 12 UTC on October 2 into the boundary conditions.

3. Synopses of Typhoon cases

3.1. Synopsis of Typhoon Ketsana (2003)

In the middle of October 2003, Typhoon Ketsana initially developed from a disturbance embedded in a reversed-oriented monsoon trough between Luzon and Guam (about 1296 km east of Luzon Island) on October 15, 2003 (**Figure 1**). The monsoon trough provided a favorable large-scale environment with high humidity and abundant low-level cyclonic vorticity for TC formation. For several days, the system remained disorganized while drifting to the west–northwest due to weak steering currents, south of the subtropical ridge. The disturbance developed into a tropical depression at 1200 UTC on October 18 (taken as formation time), and on October 19, the Japan Meteorological Agency (JMA) upgraded the depression to tropical storm, and by that time the storm had begun drifting to the northeast. Throughout the days of October 20-21, movement was slow with only weak northeasterly steering currents controlling Typhoon Ketsana, although the intensification was not so slow. With favorable outflow, Ketsana quickly intensified and was upgraded into a typhoon at 1200 UTC on October 20. After an eye formed, the slow motion continued throughout the day as did intensification. By 1200 UTC on October 21, the intensity had reached the lifetime peak intensity of 125 kt. Ketsana started weakening at 1800 UTC on October 22, with the intensity falling to 115 kt.

At 0000 UTC on October 23, Typhoon Ketsana moved slowly northeastward, but a weakening trend had set in and began to accelerate the next day, with evidence of a mass of stratocumulus cloud to the northwest of Ketsana, showing the presence of colder, drier, and more stable air. At 0000 UTC on October 24, Ketsana was beginning extratropical transition. Drier air had penetrated into the circulation and went northeastward into the westerlies with doubled forward speed.

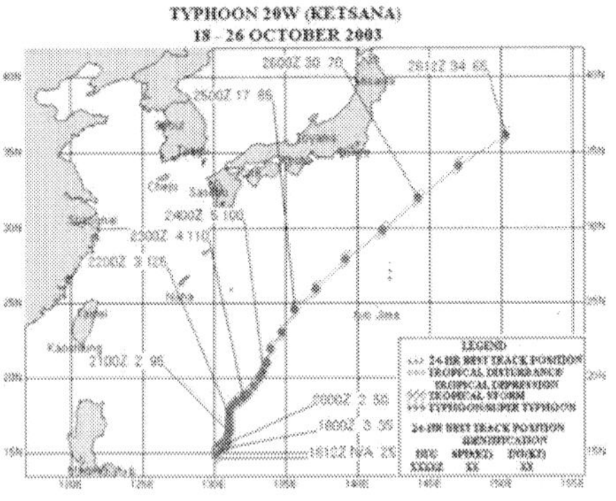

Figure 1. Best track of Typhoon Ketsana (2003) from the JTWC (adapted from Ref. [6]).

3.2. Synopsis of Typhoon Dan (1999)

Typhoon Dan (1999) first developed over the Philippine Sea at 1200 UTC on October 1, 1999 to the east of Island Luzon. The Joint Typhoon Warning Center (JTWC) issued a TC formation alert at 0230 UTC on October 2. When deep convection was seen to build over the low-level circulation center from the south near 1500 UTC on October 2, the first warning of the TC was issued by the JTWC. The system further developed into a tropical depression about 1140 km east–northeast of Manila on October 3 (taken as formation time at 1200 UTC on October 3, 1999) and then moved westward (**Figure 2**). Dan intensified very fast to a tropical storm and

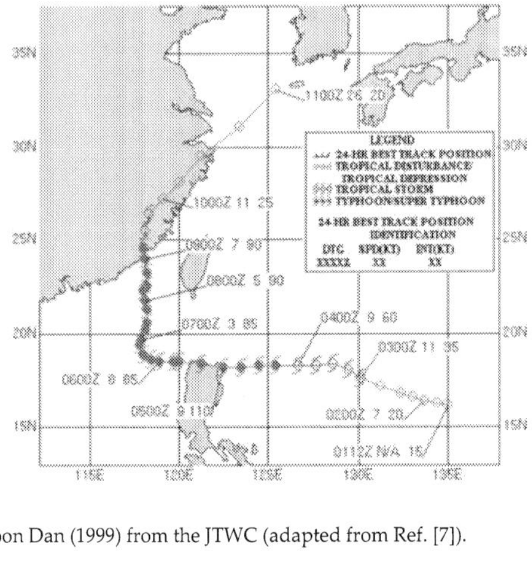

Figure 2. Best track of Typhoon Dan (1999) from the JTWC (adapted from Ref. [7]).

then a typhoon the next day. It reached a peak intensity of 110 kt (56.9 m/s) when affecting the Northern Luzon coast on October 5. TY Dan then moved over the South China Sea and weakened when it entered an increased vertical wind shear environment. It slowed down the next day and abruptly turned northward on October 7 with slight re-intensification. It eventually made a landfall near Xiamen, Fujian, China on October 9 and then weakened overland. Dan turned to the northeast and weakened to a tropical depression before it moved over the Yellow Sea later on October 10.

4. MCSs analysis

One of the characteristics of TC formation in the monsoon trough is frequent development of MCSs due to low-level convergence that enhances convection. The definition of MCS in Ref. [8] (i.e., area >1000 km^2 within the brightness isotherm −52°C) is applied. During the 48 h (1200 UTC, October 16–18) prior to Typhoon Ketsana's formation, five MCSs are observed. The first two developed on October 16 and early October 17, respectively. Later, MCS3 and MCS4 developed almost simultaneously near 1500 UTC on October 17, but then dissipated (**Figure 3**). The fifth MCS5 developed at 0600 UTC on October 18 near the low-level circulation center, and led to the formation of the typhoon 6 h later.

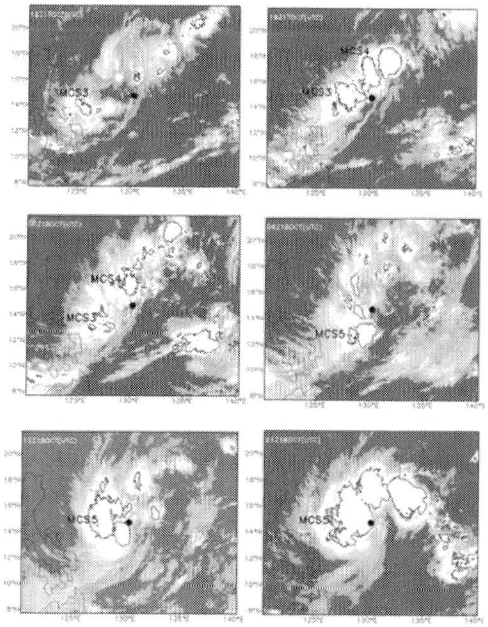

Figure 3. Six-hourly IR1 satellite images from 1200 UTC on October 17, 2003 to 1800 UTC on October 18, 2003. The contour is TB of −75°C, and the black dot is the best-track location of Ketsana's formation (adapted from Ref. [10]).

Figure 4 shows the development process of the formation of Typhoon Dan. Since October 1, 1999, there were many weak tropical convective clusters formed and maintained. At 1200 UTC on October 2, one of the cloud clusters started to develop and kept expanding to form a MCS at 0000 UTC on October 3 and then further developed to a tropical depression at 1200 UTC the same day. In contrast to Typhoon Ketsana, there was only one MCS that appeared during the formation process of Dan. While Lee et al. [1] identified that many of the TC cases with single MCS during formation were associated with easterly wave, filtered low-level winds (based on a simple running mean technique with similar low-pass effect of 3–8 days as in Ref. [9]) do not reveal wave activity during the formation of Typhoon Dan. Thus, Typhoon Dan is also classified as a typical monsoon trough formation. The focus here is the single MCS configuration associated with Typhoon Dan's formation in contrast to that of Typhoon Ketsana, with both cases embedded in similar environmental setting.

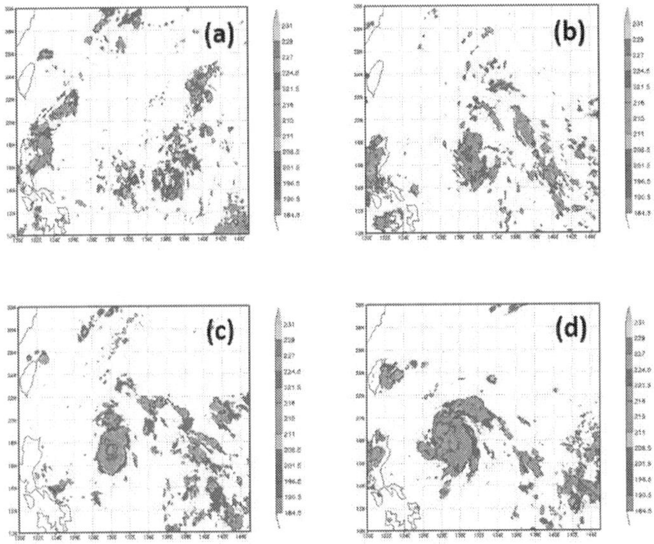

Figure 4. MCSs involved in the formation process of Typhoon Dan at (a) 0500 UTC on October 1, (b) 1200 UTC on October 2, (c) 0000 UTC on October 3, and (d) 1200 UTC on October 3, 1999. Raw pixel values have been shown. (Source: JMA GMS-5 infrared channel-1 data).

5. Model validation for Typhoon Dan

5.1. Synoptic flow

When comparing the NCEP FNL operational analysis data with the WRF simulation of Typhoon Dan in the large domain, it can be seen that the large-scale circulation has been

http://dx.doi.org/10.5772/64185

simulated well. In the FNL analysis, there are two cyclonic regions at about 130°E and 140°E (**Figure 5**). The WRF model mainly developed the incipient vortex circulation at 130°E that eventually became Typhoon Dan. Such consistency with the analysis at the low and mid (not shown) levels provides the conditions for the right timing of TC formation in the model. At the upper level, the simulated subtropical high is located north of where Typhoon Dan is developing, with the maximum high pressure center east of the Taiwan island. This is well verified by the FNL analysis (**Figure 6**), and the system provides good outflow for the formation of Typhoon Dan during its development. The simulated formation position (based on identified near-surface circulation center) and early westward motion of Typhoon Dan match with the best track very well (**Figure 7**), with the formation only a small distance east of the actual position.

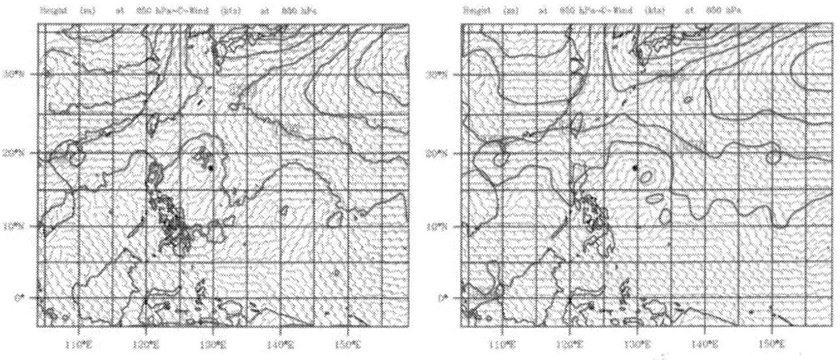

Figure 5. Simulated (left) and FNL analyses (right) of 850 hPa geopotential height (m) and wind barbs at 2000 UTC on October 3, 1999.

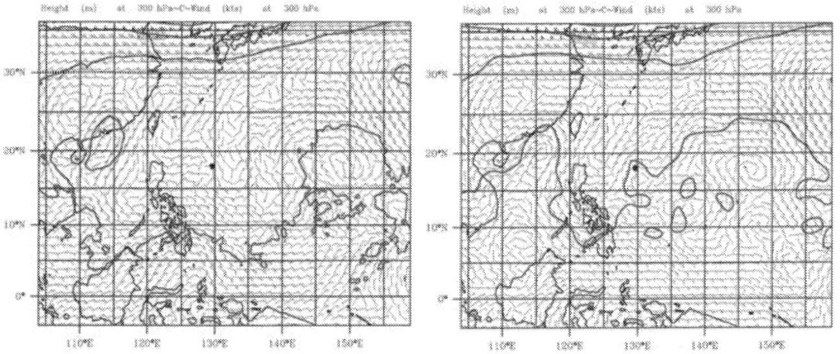

Figure 6. Simulated (left) and FNL analyses (right) of 300 hPa geopotential height (m) and wind barbs at 2000 UTC on October 3, 1999.

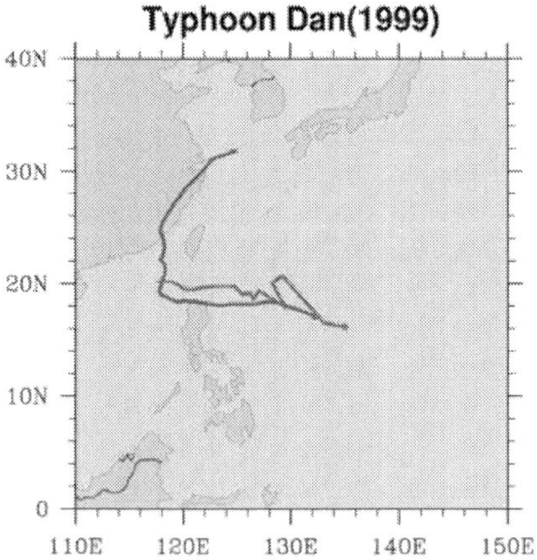

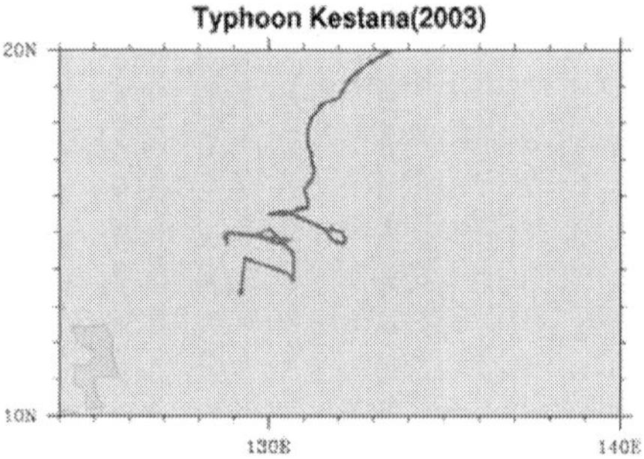

Figure 7. Simulated (red) and JMA best track (blue) of Typhoon Dan (upper) and Ketsana (lower) during 0000 UTC, October 3–6, 1999 and 1200 UTC, October 17–20, 2003 (best track has been extended after October 20, 2003 for Ketsana).

The model validation for Typhoon Ketsana has been presented in detail in Ref. [10], and thus not repeated here. The WRF model applied here is the same version as that used in Ref. [10]. The simulation of Typhoon Ketsana in this study well reproduced the reversed oriented monsoon trough in October 2003 as well as the convection episodes of all MCSs associated with Typhoon Ketsana's formation. The simulated formation position of Ketsana is southwest of the best-track position (**Figure 7**); however, the weak steering flow during the early development has been well simulated.

5.2. Evolution of MCS and intensity

In Ref. [8], the area-average observed TB from satellite images shows a minimum at around 2100 UTC on October 17, 2003 that is associated with MCS3 and MCS4, and then there is another major decrease before formation associated with MCS5 (**Figure 5b** of Ref. [10]). Simulated radar

http://dx.doi.org/10.5772/64185

reflectivity is used to examine convection activity in the model. The time series of simulated radar reflectivity has a local maximum at the same time of occurrence of MCS3 and MCS4. It then increases rapidly 6 h before Ketsana's formation, which is due to convective bursts within MCS5 and is consistent with the observed variation of the area-average TB. The simulated positions of these MCSs also match with those in satellite images: MCS3 was east of the low-level circulation center, and later MCS5 developed at a similar position (**Figure 8**). In terms of intensity, it can be seen that the simulated storm intensifies at the similar rate as observation before formation, however, becoming too intense in the rest of the simulation (**Figure 9**).

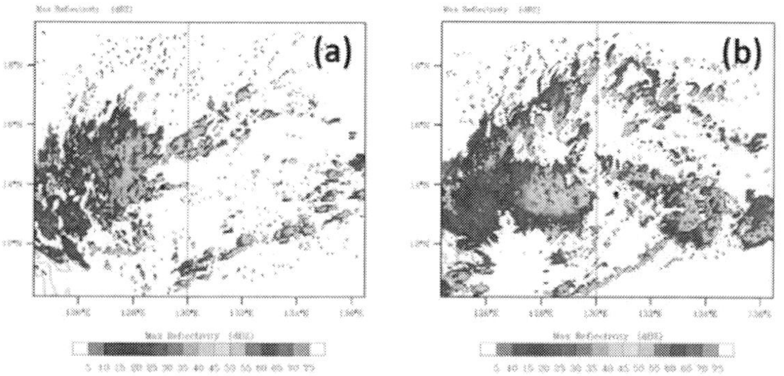

Figure 8. Simulated radar reflectivity (dbZ) of Typhoon Ketsana at (a) 1500 UTC on October 17 and (b) 0600 UTC on October 18, 1999.

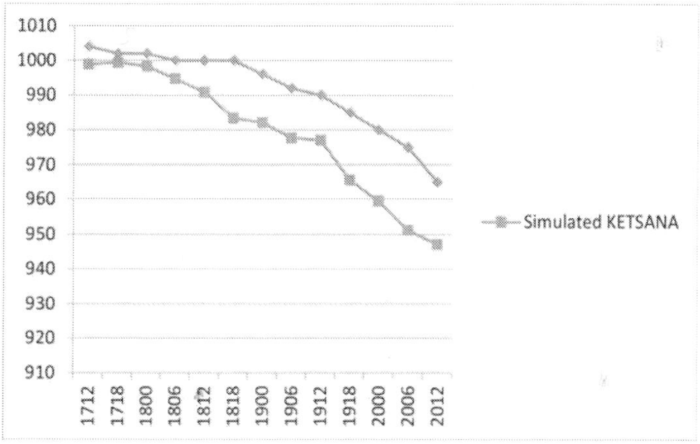

Figure 9. Simulated (red) and JMA best-track minimum surface pressure (hPa) of Typhoon Ketsana during 0000 UTC, October 17–20, 2003.

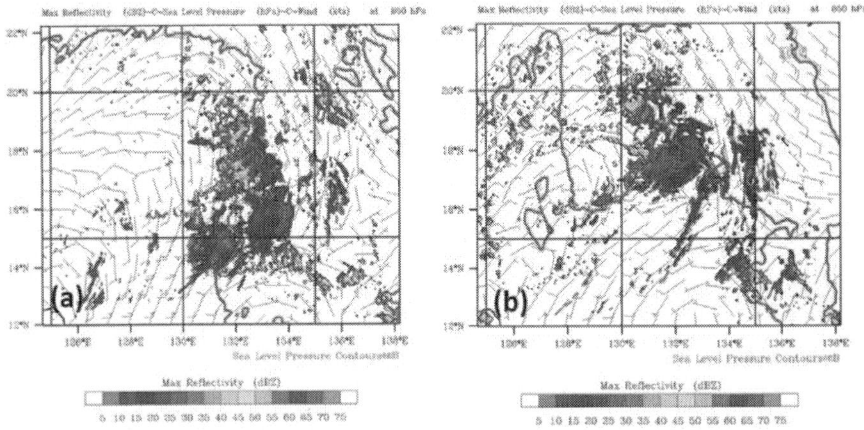

Figure 10. Simulated mean sea-level pressure, 850-hPa wind barbs, and maximum radar reflectivity at 0300 UTC on October 2 (a) and 0700 UTC on October 2, 1999 (b).

The simulated MCS activities of Typhoon Dan are also examined via the simulated radar reflectivity. In early October 2, 1999, the model simulated patches of convection on the eastern side of the broad cyclonic circulation within the monsoon trough (**Figure 10**). A few hours later, the convection organized into a MCS northeast of the circulation center. Such convection persisted when the incipient vortex moved northwestward and intensified. At the beginning of formation, there were still some weak convective cloud clusters that developed slowly until the only MCS formed. In the early formation stage, the model has not captured the initial rate of intensification very well. In the simulation, the surface pressure only started to drop from about 0000 UTC on October 3 (**Figure 11**). Nevertheless, the convection pattern that developed from the single MCS has been reproduced well in the model. From October 3, the simulated intensification rate was similar to that in the best track, and by the end of simulation the TC was more intense than that Typhoon Dan actually attained.

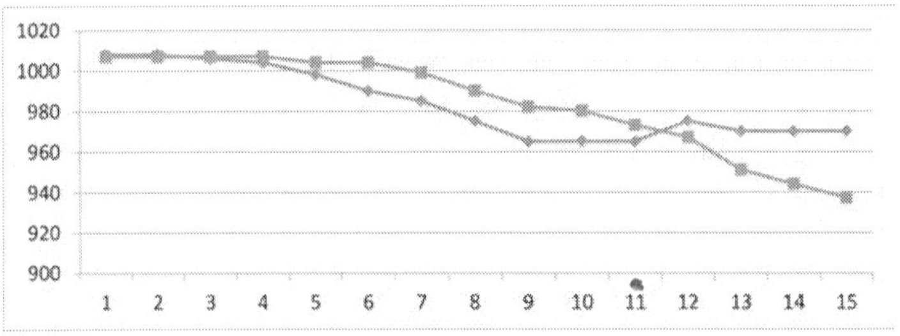

Figure 11. Six-hourly time series of simulated (red) MSLP (hPa) and that in JMA best track (blue) from 1200 UTC on October 2, 1999.

http://dx.doi.org/10.5772/64185

6. Mesoscale heating and early development in Typhoon Ketsana

6.1. Axisymmetric structure

It can be seen from the synopsis and model validation of Typhoon Ketsana that the MCS developments during its formation were highly spatially asymmetric with respect to the low-level circulation center. These MCSs would impose vorticity enhancement and heating effect to the system-scale vortex. Before such responses are analyzed, the axisymmetric structure in the simulation is first examined through the azimuthal mean of tangential and radial wind. The system center is taken as the surface circulation center.

On October 17, the axisymmetric structure of the simulated Typhoon Ketsana shows a broad low-level cyclonic circulation (**Figure 12a**). The maximum tangential wind is below 850 hPa and located between 150 and 200 km from the center. This broad circulation is consistent with the early MCS activities during that day when the first four MCSs developed more than 100 km away from the center. There was only a small region of inflow below 950 hPa, and thus the secondary circulation has not been set up in this stage.

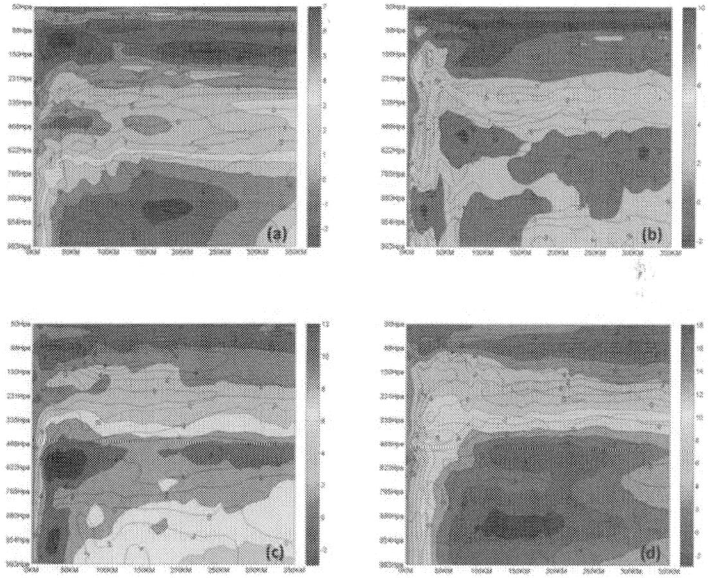

Figure 12. Azimuthal average of tangential (shaded) and radial (contour) wind (m/s) at (a) 1500 UTC on October 17, (b) 0800 UTC on October 18, (c) 1000 UTC on October 18, and (d) 0600 UTC on October 19.

On October 18, when MCS5 developed near the system center, strong low-level tangential winds started to move inward within 50 km, and at the same time extended to the midlevels (**Figure 12b, c**). The eyewall structure has been established just before the formation time of

1200 UTC on October 18. The radial winds also reveal the secondary circulation of inflow below about 800 hPa and outflow above 150 hPa.

Less than a day after the formation time, a mature axisymmetric structure is observed in the simulation. The most intense tangential winds concentrated at about 850 hPa and have been extended from 100 to 200 km from the center (**Figure 12d**). The eyewall structure is clear, and the eye has been enlarged.

6.2. Diabatic heating associated with the MCSs

It is well known that convective-type and stratiform-type rain, besides their respectively unique microphysical nature, are associated with different diabatic heating profiles [11, 12]. The latter is important to the process of vorticity enhancement and warm core during TC development. Convection is associated with deep tropospheric heating, while stratiform rain is associated with upper-level heating but possibly low-level cooling due to evaporation.

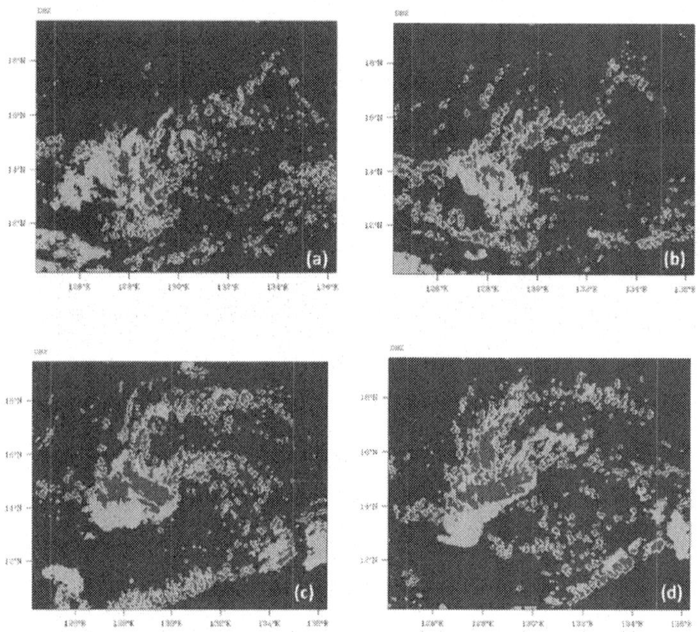

Figure 13. Convective (red) and stratiform (green) rainfall in simulated Typhoon Ketsana at (a) 1200 UTC on October 17, (b) 1500 UTC on October 17, (c) 0800 UTC on October 18, and (d) 1200 UTC on October 18, 2003.

The method in Refs. [13] and [14] is used to separate the two types of rain in Typhoon Ketsana based on the simulated radar reflectivity. The identification of the convective rainfall has further confirmed the MCS activities during the typhoon's development. On October 17, 2003, among the larger stratiform rain patches, the convective rain concentrated first to the west of the low-level circulation center (which was near 130°E, 14°N), and a few hours later to the north and northeast (**Figure 13a, b**). This is consistent with the temporal evolution of the deep convection from MCS3 to MCS4.

In early October 18, 2003, deep convection occurred again but much closer to the low-level circulation center, which is associated with MCS5. The deep convection persisted until the formation of the cyclone (**Figure 13c, d**). On the other hand, the stratiform rain areas are much larger than the convective areas at all times, which is similar to the partition of other TC cases based on the observed rainfall [15].

When the simulated diabatic heating in WRF is examined, the maximum heating is identified to locate around the midlevel, with extension from about 850 hPa up to 200 hPa. The spatial distribution of maximum heating is confined to specific locations and thus attributed to the deep convection episodes. For example, the azimuthal average of diabatic heating during October 17 shows tropospheric heating positions mostly outside the radius of 50 km, which is likely due to the early MCS activities (**Figure 14a, b**). The maximum heating of these early episodes is of the order of 10^{-3} K/s.

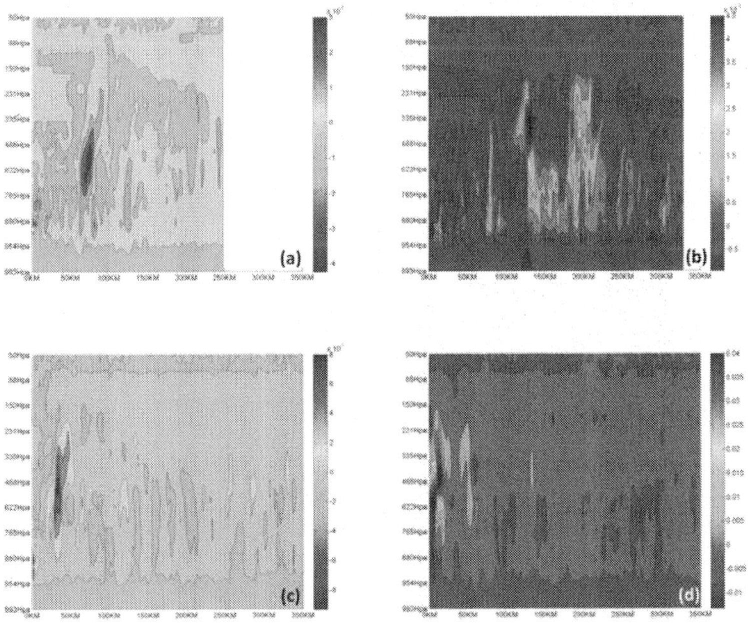

Figure 14. Azimuthal average diabatic heating in simulated Typhoon Ketsana at (a) 1600 UTC on October 17, (b) 2300 UTC on October 17, (c) 0500 UTC on October 18, and (d) 0800 UTC on October 18, 2003. Note the changes in scales.

In early October 18, it can be seen that maximum heating occurs within 50 km that is associated with the last convection episode before the formation of Ketsana (**Figure 14c**). The heating location further moves inward to the circulation center and increases in magnitude about an order higher than the previous episodes (**Figure 14d**).

6.3. Responses to diabatic heating

Reference [5] analyzed the responses of a weak TC vortex to diabatic heating based on the analytical solution of the balanced vortex model, which leads to rapid development toward the steady state with a mature warm-core thermal structure. It was found that the responses depend on the radial location of the diabatic heating. In particular, three factors of static stability, baroclinity, and inertial stability are involved in critical formulating of the responses, especially the inertial stability that usually varies substantially with the radius in a TC vortex. Diabatic heating would be most effective when it is located within the high inertial stability region, which is inside the radius of maximum wind. When the heating position is outside the radius of maximum wind, rapid development is not likely. Thus, based on these theoretical results, the convection episodes associated with the MCSs in the case of Typhoon Ketsana are increasing in heating efficiency for forming the TC. The diabatic heating in the earlier MCSs is outside the inner core with high inertial stability, which is mostly inside radius of 50 km (**Figure 15a**). This is consistent with the simulated TC that shows only slowly intensifying low-level winds in the outer region. During early October 18, when the TC incipient vortex is becoming more intense, the high inertial stability region expands slightly (**Figure 15b**). When the last convection episode associated with MCS5 occurs, the heating sits mostly in that region and is thus able to spin-up the inner-core winds rapidly (**Figure 12c**) and lead to TC formation.

Figure 15. Inertial stability in the simulated Typhoon Ketsana at (a) 0600 UTC on October 17 and (b) 0800 UTC on October 18, 2003, which correspond to **Figure 14a, d**.

Another way to examine the effects of diabatic heating, especially that with azimuthally asymmetric distribution and thus possessing eddy heat flux, is through the EP fluxes [16–20].

The EP fluxes on isentropic coordinates are calculated and analyzed for the simulated Typhoon Ketsana. The EP flux vector consists of the horizontal and vertical components, which correspond to angular momentum flux and heat flux, respectively. The divergence of the flux vector indicates the flux gradients.

During October 17, it can be seen that the EP fluxes with large divergences are mostly scattered at the large radii outside 200 km and concentrated at the lower levels (**Figure 16a**), which are associated with the heating of the earlier MCS convection episodes. The magnitudes of the low to midlevel flux vectors are larger outside 200 km compared with the inner core. The outward directions of these flux vectors indicate inward transport of angular momentum fluxes. There are also small upward components of these flux vectors, indicating inward heat fluxes.

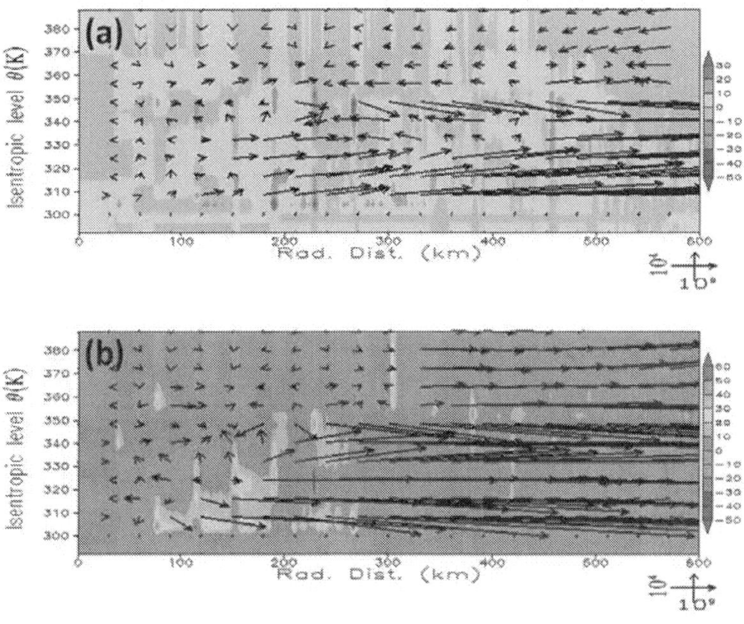

Figure 16. Divergence (contour) of the EP flux vector (arrow) in simulated Typhoon Ketsana at (a) 1200 UTC on October 17 and (b) 1800 UTC on October 18, 2003.

The locations of the largest EP flux divergence remain in the outer region throughout October 17, and then move inward during early October 18. Sometimes, the EP flux vector directions show outward transport of angular momentum in the outer low to midlevel region, which is likely due to spinning up of the outer winds. Right after the formation of the typhoon, the EP flux divergence near the inner core is still strong (**Figure 16b**). Outside radius of 100 km strong magnitudes of outward EP flux vectors indicated inward angular momentum transport again when the core vortex of the typhoon develops rapidly.

7. CAPE and the rate of TC formation

While both typhoons Dan (1999) and Ketsana (2003) formed in similar location of the WNP and the same month of year, their rates of formation were very different. Typhoon Dan developed from a disturbance to tropical depression in about 1 day; however, Typhoon Ketsana took 2 days to form from the initial MCS convection. What factors lead to these different rates of TC development? In this section, this question is examined through the CAPE consumption and recovery point of view.

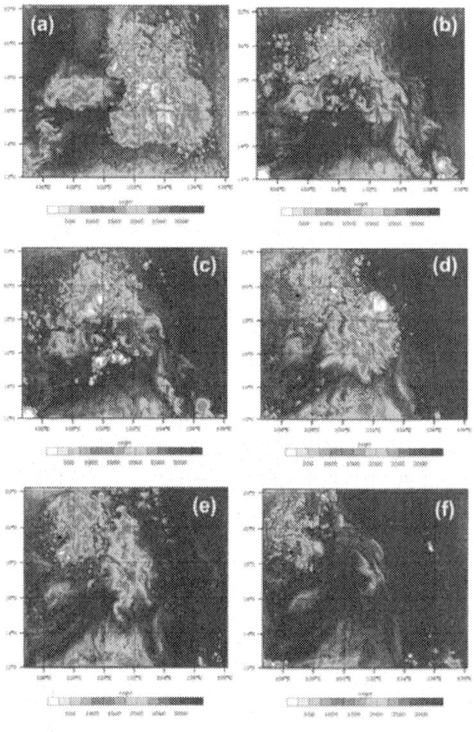

Figure 17. Simulated surface-based CAPE for Typhoon Dan at (a) 0300 UTC on October 2, (b) 1200 UTC on October 2, (c) 1500 UTC on October 2, (d) 0000 UTC on October 3 (e) 0600 UTC on October 3, and (f)1200 UTC on October 3, 1999.

When Typhoon Dan started to develop on October 2, 1999, the CAPE in the region was quite low (**Figure 17a**). The only source of high CAPE was from the southwest of the incipient circulation center. This region of high CAPE persisted throughout the day of October 2, 1999 to support the convection development of the MCS (**Figure 17b**). After the convection further developed near the circulation center, the CAPE was lowered in the region (**Figure 17c**). When the disturbance attained near-tropical low intensity at 0000 UTC on October 3, the low-level circulation center migrated to the northwest into a region still of quite low CAPE (**Figure 17d**). In other words, throughout the early development of Typhoon Dan, the CAPE with high value in that area was enough to support the development of one MCS, which was close to the low-level circulation center to establish the surface winds effectively. Thus, multiple MCSs development before formation has not been observed in this case. However, the CAPE recovered rapidly after the formation of Typhoon Dan (**Figure 17e, f**), which was good to support further intensification of the storm. This can also be identified in the area-average CAPE in **Figure 18** (the control experiment), which shows the CAPE consumption from 1200 UTC on October 2 to 0000 UTC on October 3, and then the rapid increase thereafter.

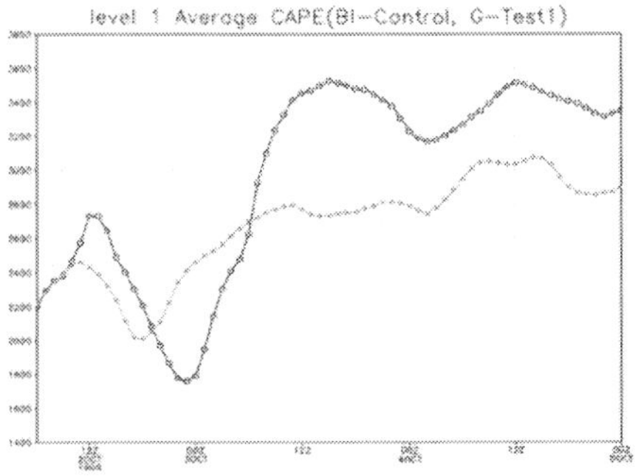

Figure 18. Area-averaged (13–17°N, 127–132°E) surface-based CAPE during formation of Typhoon Dan in the control WRF experiment (black) and sensitivity test (green).

On the other hand, the formation of Typhoon Ketsana (2003) experienced multiple cycles of CAPE consumption and recovery associated with the MCS episodes of convection. After the first two MCSs dissipated in early October 17, the CAPE on the western side of the incipient vortex was clearly lowered (**Figure 19a**). The CAPE did not recover much, and thus the two later MCSs on the same day actually developed out of only moderate values of CAPE. Their consumption of CAPE further lowered the values on the west and north sides of circulation (**Figure 19b**). At about 0600 UTC on October 18, high CAPE values entered the core region of the vortex, which supported the last (fifth) MCS to develop near the center and led to the formation of Typhoon Ketsana at 1200 UTC on October 18 (**Figure 19c, d**). The last MCS dissipated near 0000 UTC on October 19 (**Figure 19e**) that remained with low CAPE values in the core. After that the TC vortex sustained and high CAPE values reappeared associated with the eyewall convection (**Figure 19f**).

Therefore, the development of Typhoon Ketsana during the 2 days before formation was much driven by convection and thus related to the variation of CAPE in the area of development. The convection episodes associated with the MCSs spun up the winds first at the periphery and then in the core, and led to formation of the typhoon, because the four early MCSs formed and developed northwest of the core. This process was quite slow compared with the single MCS development in Typhoon Dan.

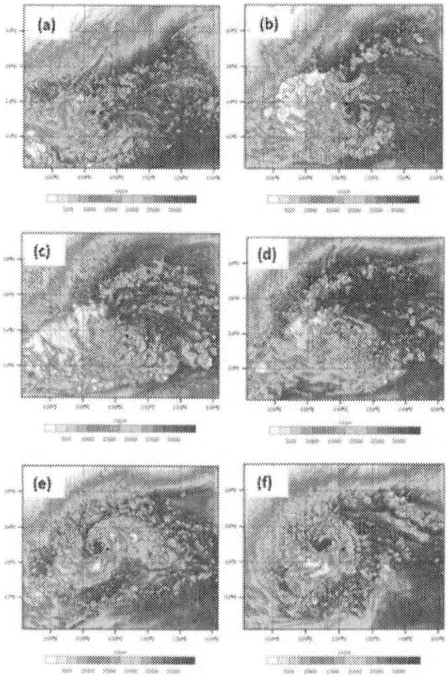

Figure 19. Simulated surface-based CAPE for Typhoon Ketsana at (a) 1200 UTC on October 17, (b) 0000 UTC on October 18, (c) 0600 UTC on October 18, (d) 1200 UTC on October 18, (e) 0000 UTC on October 19, and (f) 0700 UTC on October 19, 2003.

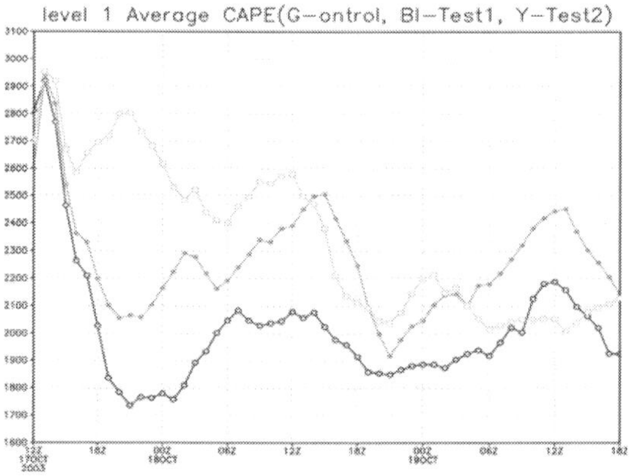

Figure 20. Area-averaged (13–17°N, 127–132°E) surface-based CAPE during formation of Typhoon Ketsana in the control WRF experiment (green), sensitivity test with strengthened MCS4 (black) and weakened MCS4 (yellow).

Since the circulation associated with Typhoon Ketsana's formation did not move much, the area-average CAPE time series also shows the consumption and recovery cycle on the day of 1200 UTC, October 17–18 (**Figure 20**, control experiment). In order to study the MCS activities of Ketsana, Lu et al. [10] designed three sensitivity experiments on the impacts of MCSs on the early intensification rate. The same set of experiments is applied here to analyze the CAPE consumption and recovery cycles. Base on satellite images, the MCS3, MCS4, and MCS5 regions are identified as (12.8–15.8°N, 127.8–130.8°E), (15.8–18.8°N, 130.8–133.8°E), and (128–178°N, 127.8–130.8°E), respectively. In the sensitivity experiment 1, the area-averaged relative humidity in the MCS3 region between 200 and 700 hPa at 1200 UTC on October 17, 2003, which was higher than that around MCS4, is assimilated into the MCS4 region within the 6-h assimilation window (similarly for later experiments), in order to strengthen the intensity of MCS4. In experiment 2, MCS4 is weakened, and only 60% of the 200–700 hPa relative humidity in the control experiment is retained and then assimilated at 1200 UTC on October 17, 2003.

The sensitivity experiments in **Figure 20** show different CAPE consumption and recovery cycles of Typhoon Ketsana. When convection in the MCS4 is stronger (black line), the CAPE drops much more than in the control experiment during late October 17. However, the recovery of CAPE is quite efficient after, and by 0600 UTC on October 18, the average CAPE value is not much lower than that in the control experiment. The early intensification rate of Ketsana is actually faster than that of the control in this experiment. On the other hand, when convection in the MCS4 is weaker (yellow line), the average CAPE consumed is less initially, then builds up slightly, and goes through another cycle before the TC formation. Although the CAPE level near formation time in this experiment is similar to that in the control, the weakened MCSs lead to a slower formation and early intensification rate.

Similar experiment as in the second sensitivity experiment of Typhoon Ketsana was performed for Typhoon Dan (green line in **Figure 18**). With a weakened MCS, CAPE consumption before Dan's formation is much smaller than that in the control. Due to the weaker convection, the early intensification rate of the simulated typhoon is also slower. After that, CAPE recovery is still quite efficient in the environment, but has not increased to the same level as in the control experiment. Therefore, from the comparison of the two typhoon cases here, it can be concluded that the rate of TC formation directly depends on how much each convection episode associated with MCS contributes to spinning up the low-level vorticity, which has been pointed out in previous studies. On the other hand, it can also be seen that how CAPE recovers in the developing TC depends on the earlier convection episodes, but not only on the large-scale environment. The control by this factor on the pace of TC's early development deserves more case studies.

8. Discussion and conclusions

It has been observed that the time for the tropical disturbances to develop into tropical storm varies with quite a large range, and is dependent on the synoptic pattern. For example, in analyzing the MCS activities during TC formations in the WNP, Lee et al. [1] found that for formations with a single MCS, some of them associated with easterly waves usually take an average of 13 h to form a TC. In contrast, for monsoon trough-related formations, there are usually more than two MCSs, and the average time to form a TC is about 1 day. Very often, more than one MCS coexist at the same time within the monsoon trough. Whereas recent studies on TC formations, such as Refs. [10, 21, 22], focused on contributions from convective-scale systems such as vortical hot towers (VHTs) and convectively induced vorticity anomalies (CVAs), their interactions with mesoscale circulations, and mechanisms to generate the core surface vortex, less discussion was on the relationship between convection patterns and development times of TCs. The numerical simulations of typhoons Dan and Ketsana here illustrate that the pace of early development of a TC depends much on the convection configuration, namely, whether it is a single MCS or interacting multiple MCSs.

It might seem to be counterintuitive that the multiple MCS convection pattern is not speeding up the TC development process due to spinning up of larger relative vorticity within more areas in the TC, but rather slowing the process down compared with the single MCS pattern. Nevertheless, this phenomenon may be explained from the point of view of competing for convective resources. Whenever deep convection occurs, the environmental CAPE is consumed. Descent of cold and dry air left by heavy precipitation is unfavorable for new convection development, which has to wait until the environmental CAPE increases again by heat and moisture fluxes from the ocean surface. When Ref. [21] analyzed the simulated formation of Hurricane Dolly (2008), similar processes were found under the TC system scale, namely, there were episodes of deep convection during the development of Hurricane Dolly, but in between there was a period when the CAPE was minimum. This interpretation is consistent with the simulation for Typhoon Ketsana, especially during the development of MCS3 and MCS4. These two MCSs developed at about the same time with a large area of convection, but did not lead to TC formation right after. The reason may be that MCS4 to the east was consuming part of the CAPE within the TC area, making the MCS3 to the west less conducive to developing more deep convection and subsequent low-level vorticity generation. In fact, MCS3 dissipated first, and the system intensification slowed down before another MCS developed in the core region.

It is interesting that this kind of CAPE argument is also applicable to the convective scales, as has been briefly discussed in Ref. [21]. Various convection processes, such as waxes and wanes, merging and splitting within each MCS are common during the early development of TC. That is, these processes represent the variability of the convective-scale VHT and CVA activities of the MCS, which is likely also due to the consumption and recovery cycles of CAPE. One point to note is that since the single-MCS convection pattern has been identified as the most common one at the end of early development, that is, for the TC to enter the intensification phase, the convection within that single MCS has to be long-lasting enough for the system to intensify to a sustainable level. In other words, there may be a hypothesis that the CAPE consumption and recovery cycles to sustain convection are more efficient under convective scales compared with mesoscales. This may be due to the fact that the areas for the surface fluxes to keep the low-level atmosphere warm and moist are simply much smaller, and/or that merging of the VTHs/CVAs is effective to sustain deep convection, especially when these systems are moving toward

http://dx.doi.org/10.5772/64185

the TC center under the system-scale convergent flow. However, these assertions have to be verified by further investigations.

In the case that merging of MCSs occurs, the resulted mesoscale vortex may lead to faster system intensification rather than the retardation as obtained in the earlier simulations [23–25]. More numerical studies on TC cases with explicit MCS merging have to be conducted to validate such hypothesis.

Author details

Kevin K. W. Cheung* and Guoping Zhang

*Address all correspondence to: kevin.cheung@mq.edu.au

Department of Environmental Sciences, Macquarie University, Sydney, New South Wales, Australia

References

[1] Lee, C.-S., K. K. W. Cheung, J. S. N. Hui, and R. L. Elsberry, 2008: Mesoscale features associated with tropical cyclone formations in the western North Pacific. Mon. Weather Rev., 136, 2006–2022.

[2] Laing, A., and J. Evans, 2011: Chapter 8, Introduction to Tropical Meteorology, 2nd edition, The Comet Program, U.S. University Corporation for Atmospheric Research.

[3] Zehr, R. M., 1992: Tropical Cyclogenesis in the Western North Pacific. PhD Dissertation. Colorado State University, Fort Collins, CO.

[4] Gray, W. M., 1998: The formation of tropical cyclones. Meteorol. Atmos. Phys., 67, 37–69.

[5] Vigh, J. L., and W. H. Schubert, 2009: Rapid development of the tropical cyclone warm core. J. Atmos. Sci., 66, 3335–3350.

[6] ATCR, 2003: Annual Tropical Cyclone Report, Joint Typhoon Warning Center, U. S. Naval Pacific Meteorology and Oceanography Center, 827 pp.

[7] ATCR, 1999: Annual Tropical Cyclone Report, Joint Typhoon Warning Center, U. S. Naval Pacific Meteorology and Oceanography Center, 214 pp.

[8] Riosalido, R., O. Carretero, F. Elizaga, and F. Martin, 1998: An experimental tool for mesoscale convective systems nowcasting. SAF Training Workshop on Nowcasting and Very Short Range Forecasting, 9–11 December, Vol. 25, Madrid, Spain 127–135.

[9] Fu, B., T. Li, M. S. Peng, and F. Weng, 2007: Analysis of tropical cyclogenesis in the western North Pacific for 2000 and 2001. Weather Forecast, 22, 763–780.

[10] Lu, X., K. K. W. Cheung, and Y. Duan, 2012: Numerical study on the formation of Typhoon Ketsana (2003). Part I: roles of the mesoscale convective systems. Mon. Weather Rev., 140, 100–120.

[11] Johnson, R. H., 1984: Partitioning tropical heat and moisture budgets into cumulus and mesoscale components: implications for cumulus parameterization. Mon. Weather Rev., 112, 1590–1601.

[12] Houze, R. A., 1989: Observed structure of mesoscale convective systems and implications for large-scale heating. Q. J. R. Meteorol. Soc., 115, 425–461.

[13] Steiner, M., and R. A. Houze, Jr., 1993: Three-dimensional validation at TRMM ground truth sites: some early results from Darwin, Australia. Preprints, 26th Conference on Radar Meteorology, Norman, Am. Meteorol. Soc., 417–420.

[14] Steiner, M., R. A. Houze, and S. E. Yuter, 1995: Climatological characterization of three-dimensional storm structure from operational radar and rain gauge data. J. Appl. Meteorol., 34, 1978–2007.

[15] Wang, Z., M. T. Montgomery, and T. J. Dunkerton, 2010: Genesis of pre-hurricane Felix (2007). Part II: warm core formation, precipitation evolution and predictability. J. Atmos. Sci., 67, 1730–1744.

[16] Molinari, J., S. Skubis, and D. Vollaro, 1995: External influences on hurricane intensity. Part III: potential vorticity evolution. J. Atmos. Sci., 52, 3593–3606.

[17] Molinari, J., D. Vollaro, F. Alsheimer, and H. E. Willoughby, 1998: Potential vorticity analysis of tropical cyclone intensification. J. Atmos. Sci., 55, 2632–2644.

[18] Chen, W., M. Takahashi, and H.-F. Graf, 2003: Interannual variations of stationary planetary wave activity in the northern winter troposphere and stratosphere and their relations to NAM and SST, J. Geophys. Res., 108, 4797.

[19] Chen, Y., and M. K. Yau, 2001: Spiral bands in a simulated hurricane. Part I: Vortex Rossby wave verification. J. Atmos. Sci., 58, 2128–2145.

[20] Martinez, Y., G. Brunet, M. K. Yau, and X. Wang, 2011: On the dynamics of concentric eyewall genesis: space-time empirical normal modes diagnosis, J. Atmos. Sci., 68, 457–476.

[21] Fang, J., and F. Zhang, 2011: Initial development and genesis of Hurricane Dolly (2008), J. Atmos. Sci., 67, 655–672.

[22] Zhang, D.-L., L. Tian, and M.-J. Yang, 2011: Genesis of Typhoon Nari (2001) from a mesoscale convective system, J. Geophys. Res., 116, D23104.

http://dx.doi.org/10.5772/64185

[23] Ritchie, E. A., and G. J. Holland, 1997: Scale interactions during the formation of Typhoon Irving, Mon. Weather Rev., 125, 1377–1396.

[24] Simpson, J., E. A. Ritchie, G. J. Holland, J. Halverson and S. Stewart, 1997: Mesoscale interactions in tropical cyclone genesis. Mon. Weather Rev., 125, 2643–2661.

[25] Ritchie, E. A., 2003: Chapter 12. Some aspects of midlevel vortex interaction in tropical cyclogenesis. Cloud systems, Hurricanes, and the Tropical Rainfall Measuring Mission (TRMM), W-K. Tao and R. Adler, eds. Meteorological Monographs, 29, 165–174.

Section 2

Tropical Cyclones and Remote Sensing

Chapter 7

Satellite Remote Sensing of Tropical Cyclones

Song Yang and Joshua Cossuth

Additional information is available at the end of the chapter

http://dx.doi.org/10.5772/64114

Abstract

This chapter provides a review on satellite remote sensing of tropical cyclones (TCs). Applications of satellite remote sensing from geostationary (GEO) and low earth orbital (LEO) platforms, especially from passive microwave (PMW) sensors, are focused on TC detection, structure, and intensity analysis as well as precipitation patterns. The impacts of satellite remote sensing on TC forecasts are discussed with respect to helping reduce the TC's track and intensity forecast errors. Finally, the multi-satellite-sensor data fusion technique is explained as the best way to automatically monitor and track the global TC's position, structure, and intensity.

Keywords: tropical cyclone, hurricane, typhoon, rainfall, intensity and track, TC monitoring and prediction, satellite remote sensing

1. Introduction

The tropical cyclone (TC) is among the most severe weather systems, with the potential for catastrophic damage to human lives, society, transportation, properties, etc. For example, Hurricane Katrina during August 23–31, 2005 with a maximum wind speed of 280 km/hr impacted most of south-east US regions and landed in the Greater New Orleans. It is the costliest hurricane in US history which killed an estimated 1245–1836 people and caused damages of $149 billion [1, 2]. Hurricane Sandy during October 22–November 2, 2012 had a maximum sustained wind speed of 185 km/hr. It led to the death of 233 people and damages of $75 billion [1, 2]. Severe TCs can attain very strong wind speeds greater than 260 km/hr and bring heavy precipitation. Most TC damage is caused by the force of its strong wind and flash flooding. TCs normally initiate over tropical oceans with Sea surface temperature (SST) >26°C to a depth of 60 meter and weak vertical wind shears [3, 4]. It can intensify rapidly under favored environmental conditions. **Figure 1** shows locations of the historical TCs and their associated intensities observed by satellite passive microwave (PMW) sensors during 1987–2012. The storm basins defined by Joint Typhoon Warning Center (JTWC) are overlaid for North Atlantic (AL), Central North Pacific (CP), Eastern North Pacific (EP), Northern Indian Ocean (IO), Southern Hemisphere (SH), and Western North Pacific (WP). The satellite-based TC distribution patterns matched very well with the JTWC's best track dataset.

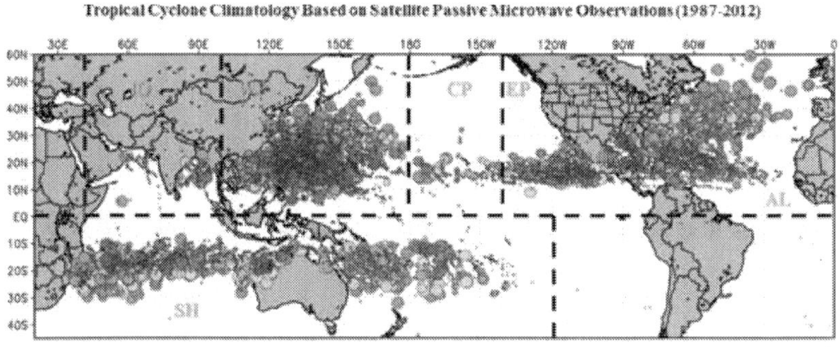

Figure 1. Tropical cyclone (TC) climatology from satellite passive microwave sensor measurements during 1987–2012. The sizes of the circles are proportional to their eye radius, while the colors show TC's intensity with warmer color for more intense TC. AL, CP, EP, IO, SH, and WP defined by Joint Typhoon Warning Center (JTWC) are used for storm basins of North Atlantic, Central North Pacific, Eastern North Pacific, Northern Indian Ocean, Southern Hemisphere and Western North Pacific, respectively.

Because a TC's life span is mostly far away from land, remote sensing—especially the satellite remote sensing—is the only way to detect and monitor global TC activities. The television infrared observation satellite (TIROS) launched on April 1, 1960 was the first experimental project of the satellite's feasibility for study of the Earth [5]. It was the first satellite used for TC monitoring and tracking [6]. The geostationary operational environmental satellite (GOES) is the key element of the United States' weather monitoring and forecasting [7]. It can be used for weather forecasting, severe storm tracking, and meteorological research. **Figure 2** shows the first images of GOES-1 at 1645 GMT on October 25, 1975. The advantages of geostationary satellites (GEO) are frequent observations over a large domain. Other countries such as China, Europe, South Korea, and Japan have their own meteorological satellite programs for improving weather monitoring and forecasting. The advanced sensors onboard operational geostationary satellites such as Japanese Himawari-8 and US upcoming GOES-R will be more powerful for providing measurements at more channels with higher spatial resolutions and more observing frequencies [7, 8].

http://dx.doi.org/10.5772/64114

Figure 2. The first image obtained from the GOES-1 satellite, October 25, 1975 1645 GMT (adapted from NOAA photo library http://www.photolib.noaa.gov/).

Low Earth orbit (LEO) polar orbital satellites provide observations over a location only twice per day. However, they can measure the 3-D meteorological conditions needed for improving weather forecasts and monitoring. Especially for TC monitoring and forecasts, conical scan PMW sensors onboard the near-circular Sun-synchronous and near-polar orbital satellites are extremely important and widely applied in operations. This conical scan pattern of PMW measurements provides for a consistent spatial resolution which is crucial in analysis of TC detection, intensity, structure, and monitoring [9, 10]. The first Special Sensor Microwave Imager (SSM/I) onboard the US defense meteorological satellite program (DMSP)-F8 was launched on June 18, 1987 [11]. Its predecessor was the scanning multichannel microwave radiometer (SMMR), which was on the Seasat and Nimbus 7 satellites launched in 1978 [12]. SSM/I was later evolved into the Special Sensor Microwave Imager Sounder (SSMIS), which provides additional measurements such as atmospheric temperature and moisture profiles [13]. The first SSMIS (F16) was launched on October 18, 2003, while the latest SSMIS (F19) was launched on April 3, 2014.

Other similar PMW sensors commonly used for TC monitoring and forecasts are the Advanced Microwave Scanning Radiometer-EOS (AMSR-E) onboard Aqua satellite and its following on the Advanced Microwave Scanning Radiometer 2 (AMSR-2) onboard the Global Change Observation Mission 1st-Water (GCOM-W1) [14, 15]; the Tropical Rainfall Measuring Mission (TRMM) Microwave Imager (TMI) was launched on November 27, 1997 and its following on Global Precipitation Measurement (GPM) Microwave Imager (GMI) was launched on February 27, 2014 [16]. The cross-scan microwave sensors such as the Advanced Microwave Sounding Unit (AMSU) onboard NOAA-15 satellite launched on May 13,1998 and its next generation Advanced Technology Microwave Sounder (ATMS) onboard the National Polar-orbiting Partnership (Suomi-NPP) launched on October 28, 2011 have also been applied in TC monitoring and intensity measurements because of their superior spatial resolution [17–19]. These PMW sensors are critical in satellite-based rainfall retrievals [20] that are utilized in TC precipitation observations and forecasts [21].

Section 2 discusses TC detections from satellite observations, while Sections 3 and 4 describe analysis of TC structure and intensity with satellite remote sensing. Section 5 is about TC rainfall while Section 6 discusses the impacts of satellite remote sensing on improving TC forecasts. Global TC monitoring and tracking with multi-satellite sensors are discussed in Section 7. Section 8 presents a brief summary of history, applications, and impacts of satellite remote sensing on TC structure, intensity, forecast, and monitoring.

2. Identification of TC using satellite measurements

The TC is a highly organized weather system which can be easily detected by satellite observations. The deep clouds of TCs are shown in GEO Infrared (IR) measurements as low temperatures compared with cloud-free or thin cloudy areas because the IR sensors can only detect the cloud top without measuring anything underneath it, while the satellite Visible (VIS) observations show high values of reflectivity because of the clouds' high albedo. The GEO satellite is an ideal platform to detect TC activities because of its frequent measurements every 30 minutes, and even 10 minutes for next generation GEO sensors such as the Japanese Himawari-8 and the upcoming US GOES-R sensors. Due to the unique TC eyewall feature, it is easy to identify the position and lifecycle of a well-developed TC because of its apparent eye location where cloud-free or thin clouds exist. **Figure 3** presents an example of observations over West Pacific from Himawari-8 IR. It is apparent that two typhoons (Goni and Atsani) are clearly displayed at 1800 UTC August 18, 2015.

Figure 3. Japanese satellite Himawari-8 IR image of Typhoons Goni and Atsani at 1800 UTC Aug 18, 2015 (adapted from the Joint Typhoon Warning Center (JTWC) at http://www.usno.navy.mil/JTWC).

http://dx.doi.org/10.5772/64114

There are limitations in satellite IR/VIS observations at early stages of storm development when its center is not obvious or overcast. The widely used analysis method for satellite remote sensing of TC is called the Dvorak technique [22]. It is based on combinations of the enhanced IR image analysis and detection of cloud shapes at different stages of storm development. Applications of the LEO satellite PMW remote sensing lead to large improvements in TC's detection in terms of structure, location, and intensity because of the PMW sensor's superior spatial resolution and sensitivity of cloud vertical profiles due to its capability to penetrate clouds. The PMW imagery at high-frequency channels shows a depletion of brightness temperatures (T_Bs) over deep cloud areas where the scattering effects of frozen hydrometeors on microwave radiance exist. **Figure 4** presents an example of the multi-sensor combined four panels of IR/VIS and PMW high-frequency channel imagery for Typhoon Mindulle over west Pacific at 1800Z of June 26, 2004 from Naval Research Laboratory-Monterey (NRL-MRY) TC web page at http://www.nrlmry.navy.mil/TC.html. This kind of multi-panel display is used for a better analysis of the TC intensity, position, and structure. The upper-left panel is the GOES IR image, while the upper-right panel is the enhanced IR image (VIS imagery is used in daytime when it is available) with the Dvorak method for improved TC analysis. For this TC case, the overcast blocks a good view of the TC center from IR observations. The enhanced image is better in displaying the potential TC center, but is still not able to show the TC eyewall and

Figure 4. Images of Hurricane Mindulle at 1800UTC June 26, 2004 from GOES-9 IR and AQUA AMSR-E. Top left panel: GOES-9 IR; top right panel: enhanced TC IR analysis known as the Dvorak hurricane curve for tropical cyclone classification; bottom left panel: AMSR-E 89H GHz; and bottom right pane: AMSR-E composite (adapted from the NRLMRY-TC web page at http://www.nrlmry.gov/TC.html).

center. The bottom-left panel is an image of the AMSR-E 89H GHz channel, which clearly shows positions of the TC eyewall and center as well as convection cells. The bottom-right panel is a composite image of the PMW polarization corrected temperature (PCT) in red, 85 GHz or 89 GHz vertical polarization in green, and horizontal polarization in blue [23]. It can provide additional information of the TC cloud patterns important to TC's overall structure and organization, especially for a potential low-level circulation center. Therefore, this multi-sensor-panel imagery can be applied for better analysis of these important TC characteristics such as eyewall development, formation of concentric eyewall, convective zones and cells, rain band formation, central dense overcast, shear, and low-level center.

3. TC structure analysis

Soon after the first satellite weather observations became available, routine TC intensity studies from this imagery began [23]. Determining the TC's center position was the only reliably produced operational use of TC satellite analysis until the influential work of Dvorak [22] in 1972. Dvorak developed an empirical model to diagnose TC intensity based on cloud organization in the visible channel. Factors that affected strength of the storm were accounted by structural features such as magnitude of the brightness, temperatures, and curvature of the banding or distortion of the cloud pattern. A flowchart of rules in the Dvorak method was created to consistently generate subjective estimates of TC intensity from the structural cues in satellite imagery. Dvorak refined his method with time to incorporate additional rules and constrictions as well as use of infrared channels, aiming to improve the accuracy and reduce subjectivity of the technique.

The Dvorak technique continues as a critical part of operational analysis today. It is still the most common method to diagnose TC intensity, with further attempts at intensity analysis heavily borrowing for this work's legacy. Some efforts continue to hone the technique's accuracy, such as updating the wind-pressure relationship used to define the intensity [24]. The application and customization of Dvorak analysis worldwide described in [23] help demonstrate its versatility and robustness as a tool for TC forecasting. Efforts to make the process more objective led to the creation of a fully automated TC intensity analysis [25], which itself continues to have updates as the Advanced Dvorak Technique (ADT) [26].

Beyond the Dvorak technique, there are several other methods to diagnose TC structure and intensity from infrared and visible channels. The correlations between infrared T_Bs and operational storm size metrics (i.e., radius of maximum winds (RMW), and radii of the 34-, 50-, and 64-kt winds) were presented in 2007 [27]. A similar use of infrared imagery to resolve TC sizes, as defined by the radius of the 5-kt 850 hPa wind, that relate to the TC lifecycle were studied [28]. A statistical analysis of the distribution of temperatures with respect to TC axisymmetry (the deviation angle variance technique) continues to show great promise in not only diagnosing structure, but also helping with TC centering, genesis, and intensity [29]. Another potential relationship between TC intensity and structure changes can be seen by differencing the GEO water vapor and infrared channels [30]. This methodology leverages information in each channel (due to the weighting function representing different altitudes and chemical profiles) to emphasize specific structural features such as overshooting convective tops. Finally, there are efforts to relate the rotational speed of IR and visible cloud tops about the TC center to the intensity of major typhoons in the northwestern Pacific basin [31].

Besides the traditional infrared and visible channels, other remote sensing frequencies have proven useful for observing TCs. The use of satellite active microwave radars (scatterometers) to estimate storm size and intensity via surface wind analysis has been very beneficial to operational efforts at the National Hurricane Center (NHC) [32]. Scatterometers transmit pulses that bounce off the ocean surface; the backscatter's variability due to wind roughening enables retrievals of the surface wind vector estimates. QuikSCAT scatterometer data were used to derive climatology of storm sizes at the 23 kt radius as well as an outer radius using a wind structure model [33]. Similar studies were done to create a QuikSCAT climatology of storm sizes for the 34-kt radius and outer-core strength (OCS) intensity [34–36]. TC structure and intensity are also investigated using other satellite sensor datasets such as the TRMM precipitation radar (PR) and TMI in tandem with lightning flash density to compare differences in frequency thresholds between regions of the TC [37]. The synthetic aperture radar (SAR) was used to visualize extremely small mesoscale details of TC and subjectively catalogue characteristics of the eye, including spatial area, shape, and wavenumber [38]. SAR TC retrievals are currently poorly tied to physical processes and the radar retrievals from space occur infrequently, while passive microwave can directly characterize TC structure by penetrating non-raining cloud tops, unlike in the visible and infrared channels [39].

While the ultimate purpose and mechanism of the TC eye formation remain uncertain, there are many characteristic modes associated with TC eye. In general, due to dynamical considerations of the eyewall, the TC eye tends to get smaller. Eventually, a new outer eyewall may form and produce an outer eye that encompasses the previous smaller eye. In rare cases of an extremely intense and compact TC, a "pinhole" eye will form [27]. In some other cases, a large and stable eye with multiple embedded meso-vorticies can form [40, 41]. The TC eye shape is not always circular, as polygonal eyewalls seem to be due to vorticity instability [42, 43]. Recent observations show eye-like features developing in the lower troposphere before being observed in the upper troposphere [44].

The strongest vertical motion in TCs was found just inward of the RMW [45, 46]. In addition, the eyewall tended to slope outward with height due to the TC warm core. The TC intensity and eyewall slope relationship shows a great deal of case-to-case variability [47]. Due to this slope, the updraft itself was also tilted so that most of the falling hydrometeors in the eyewall (and thus the radar reflectivity maximum) lie outside the RMW [46]. The high-resolution airborne Doppler radar was recently used to update and extend these results [48, 49]. Regions of the TC core, defined by normalization with respect to the RMW, are shown to exhibit modes of radar reflectivity, convergence/divergence, and vorticity that correspond to the previously cited work; particularly, there is an outer peak in upper-level divergence and low-level convergence that occurs in the vicinity of secondary eyewall formation.

The nature of TC convection occurring in spiral bands was not known until their first observations on radar [50]. However, the TC spiral band was quantitatively and qualitatively characterized shortly thereafter [51, 52]. A logarithmic spiral based on radar observations was introduced to start at an inner radial circle (i.e., the inner-core or eyewall) rather than the center itself [51]. An analysis of the logarithmic spiral length has been quite useful in diagnosing intensity [22]. The idea that low-latitude TCs have lower spiral crossing angles than higher latitude storms possibly due to storm motion was also introduced [51]. The TC spiral bands propagate outward [52], while some rain bands actually propagate inward when taking into account storm motion [53]. In general, there may be three types of spiral bands based on movement: stationary (non-propagating), apparent propagation (stationary with respect to the TC center), and intrinsic propagation [50].

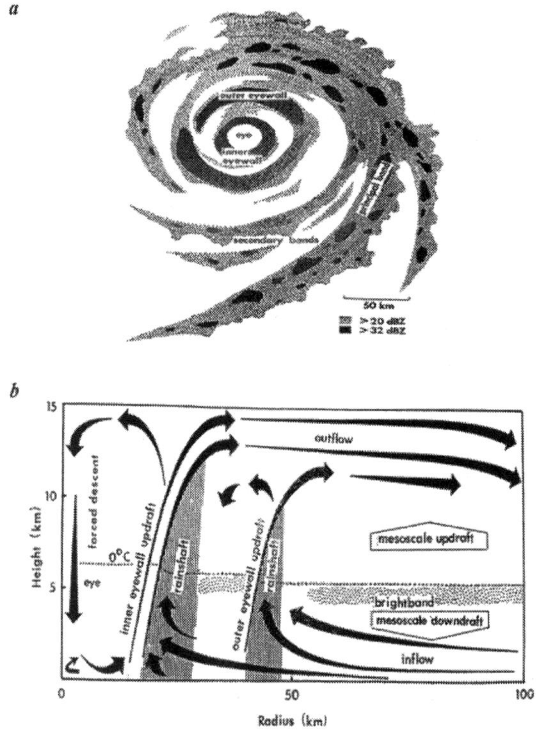

Figure 5. Two schematics of TC structures from [109]. (a) Horizontal cross-section of structural features as presented by radar. (b) Vertical cross-section of the same structural features and their relation to the secondary circulation described in Eliassen [110] and Shapiro and Willoughby [111] (adapted from Willoughby (1995)).

These structural characteristics are summarized in **Figure 5** for a well-organized double eyewall TC. The inner eyewall, outer eyewall, principal convective band, and secondary convective band are clearly presented with radar reflectivity (**Figure 5a**). The stratiform precipitation occurs largely in the moat areas. These unique TC features are clearly captured by the PMW sensor's measurements at high-frequency channels as shown in **Figure 4**. The TC's tilted updraft at the eyewall, forced descent air at the eye, lower level inflow and upper level outflow, brightband associated with stratiform precipitation as well as mesoscale updraft (downdraft) above (below) the brightband are demonstrated in **Figure 5b**. Detailed TC vertical temperature profiles can also be observed by PMW sensors. **Figure 6** shows cross section of the AMSU-retrieved temperature anomalies through hurricane Bonnie at 1200 UTC August 25, 1998. The TC warm core near 250 hPa and the vertical temperature profiles match well with observations. Thus, this unique warm core feature can be applied for TC intensity estimates discussed in the next section.

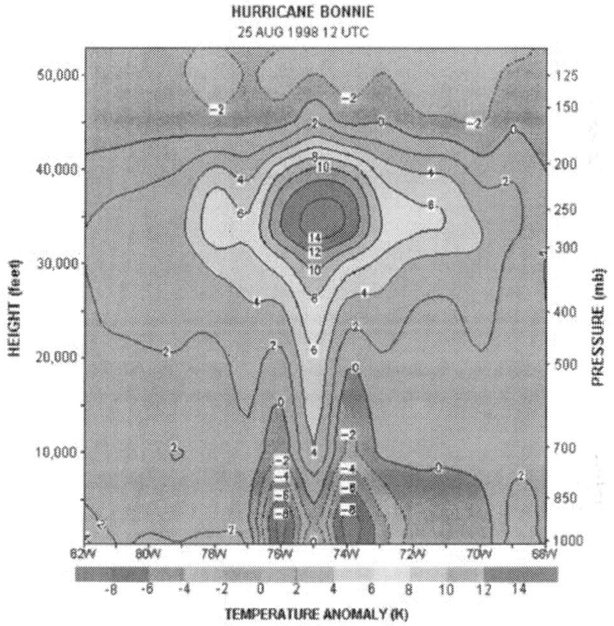

Figure 6. Cross section of temperature anomalies through Hurricane Bonnie at 1200 UTC 25 Aug 1998 retrieved from AMSU data (adapted from Kidder et al. (2010). ©American Meteorological Society. Used with permission).

4. TC intensity estimation

Generally, there are two types of available polar-orbiting microwave sensors: imagers and sounders. The microwave imagers typically consist of sensors with frequencies that measure surface properties as well as organization of various water phases in the atmosphere. The microwave sounders aim to provide profiles of atmospheric thermal structure and moisture estimates. Depending on the mission goals of a particular sensor, there can be overlap between available channels of a microwave imager and sounder.

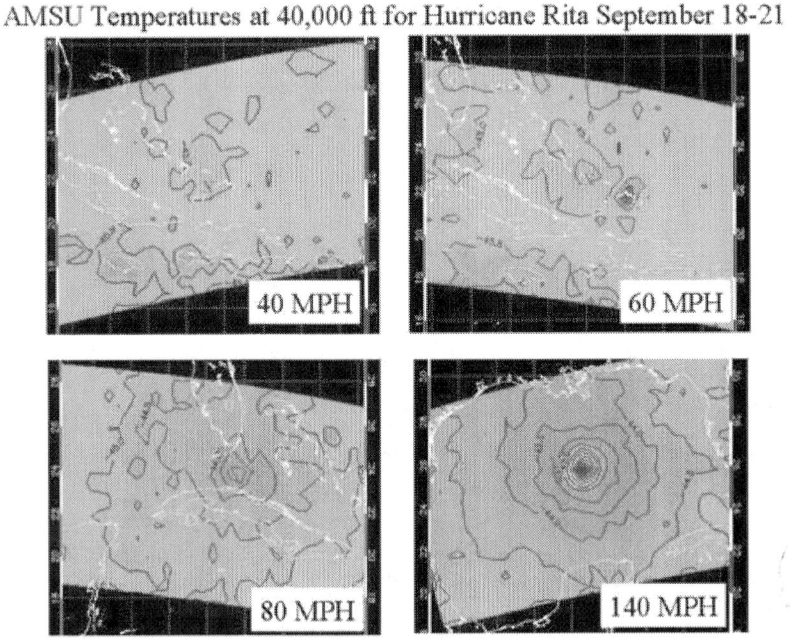

Figure 7. A multi-panel comparison of the AMSU temperature structure at the upper levels of Atlantic Hurricane Rita (2005). "X" is the TC center position. The strength of the temperature anomalies represents deepening of the warm core structure and corresponds well with increasing intensity (adapted from UW Madison/CIMSS: http://tropic.ssec.wisc.edu/real-time/amsu/).

Attempts to diagnose TC structure and intensity from microwave sounders occurred shortly after the first sounder was launched [54]. Although the coarse resolution of sounders has traditionally created an analysis barrier due to smoothing over the storm features, the more advanced sensors such as AMSU are starting to resolve the magnitude of thermal anomalies as well as core/eye size more faithfully [19, 55]. **Figure 7** presents the AMSU-retrieved temperature anomaly distributions at 4000 ft at different stages of Hurricane Rita during 18–21 September, 2015. It clearly shows increased amplitude of the unique TC upper level warm core feature corresponding well with intensification of the hurricane intensity. Multiple linear regressions of the AMSU channels also can estimate features such as maximum sustained wind (MSW), minimum sea level pressure (MSLP), and wind radii at the 34-, 50-, and 64-kt thresholds [56]. Use of AMSU data as part of an ensemble shows great promise for more accurate TC structure retrieval; a combination of the ADT and two different AMSU intensity estimates makes up the satellite consensus (SATCON) method of TC intensity estimation [57], which has

http://dx.doi.org/10.5772/64114

the highest skill of all satellite-based intensity estimation methods [58]. A comparison of the TC sustained 1 minute wind estimates from different techniques is displayed for an example storm in **Figure 8**. It demonstrates that the PMW sensor-based measurements can be used to accurately estimate the TC intensity. Although every method with individual PMW sensor in general agrees with each other, differences are still obvious. Results indicate that SATCON performs well against the TC's best track dataset. These techniques have been continuously evolved to create a better sounder TC intensity algorithm with new sensors such as SSMIS and ATMS, which have improved spatial resolution to depict the TC's warm core [18].

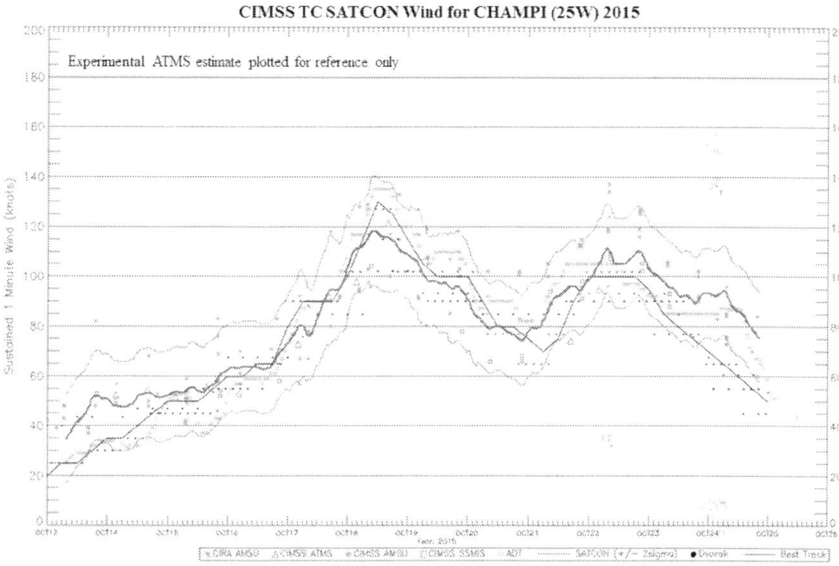

Figure 8. Satellite consensus (SATCON) intensity analysis of Typhoon Champi (2015) in the West Pacific. The solid black line shows the best track intensity from JTWC, the black dots show the subjective Dvorak satellite estimates, and all the other plots show objective satellite-based estimates (adapted from Derrick Herndon and CIMSS, http://tropic.ssec.wisc.edu/real-time/satcon/).

To contrast, microwave imagers have a more recent appearance in TC analysis, with perhaps more potential for added value. The near real-time access to digital microwave imagery was the largest impediment near the turn of the millennium [39], in which authors describe efforts at NRL-MRY to provide near real-time access to high-resolution TC images, and to support a temporal return frequency favorable for operational TC forecasting. A more detailed discussion about the utility of microwave imagers indicated that despite the ability to create the PMW-derived physical quantities (e.g., sea surface wind magnitude, precipitable water, and cloud liquid water) using multiple frequencies which utilize different spatial resolutions, they smooth over important structural features of the TC [59]. These features include a $_{TB}$ depression at the high frequency channel (85, 89 or 91 GHz) due to ice scattering and lower frequencies such as 37 GHz principally showing liquid hydrometeor emissions near and below the freezing level. Both of these previously mentioned channels are measured at horizontal (H) and vertical (V) polarizations. Near the interface of the outer TC and the environment, interpretations at either polarization become muddled due to multiple competing influences (e.g., water vapor, cloud water, and sea surface). The polarization correction temperature (PCT) can improve the representation of atmospheric features, allowing them to stand out from surface background [60].

Despite its relatively new arrival, some progresses are apparent in using microwave imagers to examine tropical cyclones. For example, the NHC extensively uses microwave imagery to better locate a TC center and subjectively diagnose changes in structure [32]. The Morphed Integrated Microwave Imagery at CIMSS (MIMIC), a technique to create "morphed" animations of passive microwave imagery using an advection function between satellite passes, was introduced to allow a visually appealing depiction of TC structure changes [61]. Other studies also revealed relationship between microwave imager data to TC intensity [62, 63]. The microwave data have been used to improve TC intensity estimates through early detection of a forming eyewall [64], while a color composite of the H-pol, V-pol, and PCT data at 37 GHz developed at NRL-MRY has shown particular promise in diagnosing TC inner core formation [39]. A symmetrical and closed T_B threshold ("cyan ring") was applied to predict the TC onset rapid intensification [44].

Some efforts focused on cataloging TCs through an extended climatology of microwave imagers. The microwave data interpolated onto an 8-km grid in the hurricane satellite (HURSAT) archive was created in 2008 [65]. The HURSAT-microwave consists of data from the SSM/I platforms between 1987 and 2009, using global best track data from the International Best Track Archive for Climate Stewardship (IBTrACS) to search for TCs [24]. Based on this dataset, the TCs composited by their intensification rate and environmental wind shear were analyzed to compare microwave signatures during different intensity regimes [66]. Recently, a new study on eyewall size estimates using the HURSAT-microwave data compared to the aircraft reconnaissance measurements demonstrates the similarity of in-situ and satellite-derived structural profiles [67]. A more advanced TC PMW TB database at 1 km spatial resolution has been developed at NRL-MRY from all available PMW sensors in 1987–2012 with more consistent inter-calibrated TBs at 89 GHz, better eye fixing, and high quality interpolation scheme [68, 69].

5. TC precipitation

One of the typical phenomena of TC activities is heavy rainfall, which is also one of the most significant impacts of TCs. The tremendous precipitation from TCs often leads to loss of lives and properties. Flooding from landfall TCs over United States is the leading cause of death related to severe storms [70]. However, TC precipitation can also bring in major economic benefits to areas surrounding its path. Based on the long-term TRMM rainfall measurements, over 84% of continental convective rainfall is contributed from rain intensity > 5 mm h^{-1} [71].

Analysis of the numerical weather prediction (NWP) model rainfall forecasts indicates TC precipitation could contribute 15–17% of the total annual rainfall over broad latitude zones [72]. TC rainfall can contribute up to 15% of total precipitation over a hurricane season in Carolinas of United States [73]. Therefore, even precipitation from one TC activities could ease the stress of drought over some areas. A good review of TC rainfall's structure, intensity, and forecasts was recently reported [21].

Rain retrievals from advanced algorithms based on PMW measurements over ocean have been proven accurate and reliable [74–77]. The satellite-derived instantaneous rain patterns over TCs clearly show the heavy rainfall is normally located in TC eyewall and spiral convective areas. In general, the intensity and pattern of TC precipitation are strongly associated with the TC intensity and radial distance to eyewall [78]. The maximum rainfall appears in the TC eyewall around less than 50 km radii and the rainfall intensity rapidly reduces with increase of its radii. Rain intensity is about 13 mm h^{-1} for major TCs (category 3–5), 7 mm h^{-1} for minor TCs (category 1–2), and 3 mm h^{-1} for tropical storms. By the radii of 300–350 km, rain intensity for all kinds of TCs is almost same.

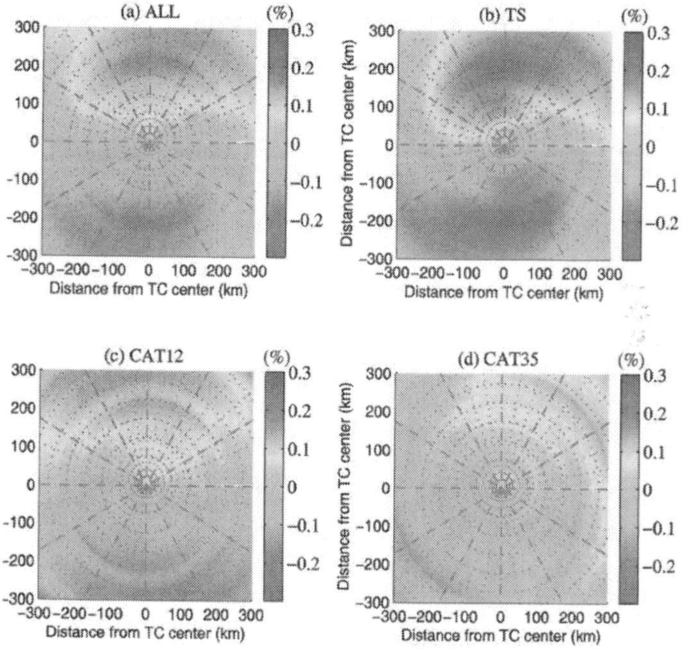

Figure 9. Rainfall asymmetry calculated in 10-km rings around the storm center, as a function of storm intensity: (a) 2121 TC observations (total distribution), (b) TS, (c) CAT12, (d) CAT35. The storm motion vector is aligned with the positive *y* axis. The color scale indicates the amplitude of the normalized asymmetry. Red corresponds to the maximum positive anomaly and blue to the minimum rainfall within the storm (adapted from Lonfat et al. (2004). ©American Meteorological Society. Used with permission).

The asymmetry of TC precipitation is a prominent feature. It shows different characteristics depending on what matric is applied. **Figure 9** presents the TC rainfall asymmetry patterns relative to its motion direction as a function of storm intensity based on 3 years' TRMM TMI rain retrievals [78]. For all storms and tropical storms, their maximum rain intensity is in the front quadrants of TC movement. The location of maximum rain intensity shifts from the front-left for CAT1-2 to front-right quadrants for CAT3-5. Thus, the asymmetry of TC rainfall is linked with the TC intensity, especially for strong TCs. In addition, the asymmetry has a property of strong dependence on TC geographic locations. Maximum rainfall appears in front quadrants over WP while in front-right quadrants over AT. Maximum rainfall shifts to the front-left quadrants over SH. Over EP and IO, it is located in the front quadrants with a cyclonical pattern.

A recent study based on TRMM rain datasets for landfall TCs over different parts of China presents various rainfall patterns relative to TC's motion at different times of landfall [79]. Maximum rainfall is located in the left quadrants for TCs landed in Guangdong province and Taiwan, while in the front-left quadrants for TCs landed in Hainan and Fujian provinces. Maximum rainfall is generally located in the back-right quadrants for TCs landed in Zhejiang province. However, maximum rainfall is generally positioned in the front quadrants of TCs relative to its vertical wind shear vector (**Figure 10**), although there is still a slight difference in rainfall distribution between different areas. This feature is an important finding because it has a potential application for improving TC rainfall forecasts.

What are the possible causes for the asymmetric distribution of TC precipitation? Several known factors are associated with this asymmetry feature, such as the advection of planetary vorticity, vertical wind shear, and friction-induced boundary layer convergence [80, 81]. The maximum rainfall in the front-quadrants of TC motion indicates the role of friction-induced boundary layer convergence. Higher TC moving speed leads to strong rain intensity in its front quadrants [78], while its dependence on geographic locations shows importance of the TC's ambient wind influence. The improved consistence of rainfall asymmetry relative to its vertical wind shear for the landfall TCs over China further indicates the role of interaction between TC and its environmental forcing. However, this feature needs more verification studies over other TC basins and its connections to amplitudes of the vertical wind shear.

A better prediction of the TC rainfall distribution is the ultimate goal of efforts in mitigating TC's rainfall impacts on society, life, and property. Several methods have been developed for operational TC rainfall forecasts. A popular one is the Tropical Rainfall Potential (TRaP) which is based on the satellite rainfall estimates, persistence of TC intensity, and satellite-derived wind vector [19, 82].

$$\text{TRaP} = R_{\text{av}} \cdot D \cdot V^{-1} \tag{1}$$

where R_{av} is the mean rainrate along a line in the direction of TC motion, D is the distance of that line across the TC rain area, and V is the TC's actual speed. Accurate satellite rain retrievals and satellite wind vectors as well as its easy implementation have made this method popular

http://dx.doi.org/10.5772/64114

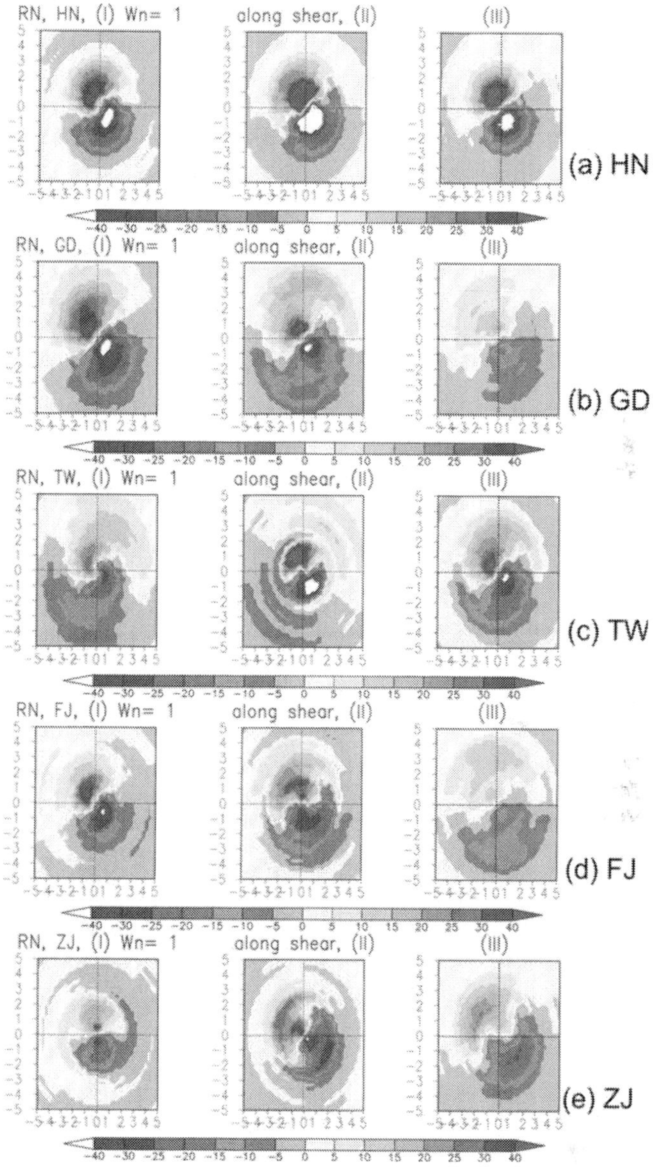

Figure 10. The wavenumber 1 rainfall asymmetry (mm) relative to the storm vertical wind shear. The shear vector is aligned with the positive y axis (upward). The x and y axes are distance (°) from the TC center (origins). Stage (I) is 24 hr prior to, stage (II) is at the time of, and stage (III) is 24 hr after landfall. The color scale indicates the amplitude of the asymmetry relative to the storm motion (adapted from Yu et al. (2014). ©American Meteorological Society. Used with permission).

in operations. A variation of this method called the areal TRaP has been introduced to graphic view of TC precipitation horizontal distributions [83, 84], with three correct assumptions of TC track forecast, satellite rain estimates, and persistent spatial pattern of rainrates relative to the TC center. TRaP is normally valid for short forecasts of less than 24 hr. However, there are limitations associated with TRaP because of no considerations on changes of TC intensity and the TC's vertical wind shear conditions.

Another method is the rainfall climatology and persistence (R-CLIPER) model which is a parametric model utilized with the TC rainfall climatology from satellite measurements [85]. This method assumes a circularly symmetric distribution of rainfall and its rainfall is translated in time. Although it accounts for TC intensity and moving speed, it does not include the TC's unique asymmetry rainfall patterns. An improved method called the parametric hurricane rainfall model (PHRaM) was introduced to incorporate with the TC rainfall asymmetry feature by including the azimuthal Fourier decomposition for shear and a term indicating the topographical uplift [86]. Results show that PHRaM is improved significantly compared with the standard R-CLIPER. A new parametric model was recently developed for including more factors such as TC motion speed and intensity, vertical wind shear, and typical features of the TC boundary layer [87].

All the above-mentioned methods have one common assumption of the correct satellite rainfall retrievals. Although satellite-derived precipitation from PMW sensors is very accurate over ocean, there are still relatively large errors over land [75, 88]. Some discrepancies exist among satellite rain datasets, especially with different satellite sensors and retrieval algorithms. In order to minimize errors from different rain retrieval algorithms for different sensors, the physical-based inversion rain algorithm (GPROF) used in TRMM and GPM is also applied for other PMW sensors [77]. Thus, precipitation from different PMW sensors will be more consistent. The new NASA integrated multi-satellite retrievals for GPM (IMERG) is based on these consistent PMW rain retrievals and calibrated IR-based rainfall so that it will produce a higher quality precipitation data [20]. The high quality satellite rainfall will be used to generate a better TC rain climatology. In addition, precipitation has a strong diurnal cycle property [89–91] and there is also a clear diurnal feature in TC lifecycle [92]. These diurnal properties could also be utilized in improving TC rain forecasts in the near future. However, the best TC rain forecast should come from future advanced cloud-revolving models which could predict not only TC intensity and track, but also rain distributions at different spatial scales.

6. Impacts of satellite remote sensing on TC forecasts

The NWP models are used at major weather operation centers to provide regular 7-day weather forecasts. The prediction skills have been consistently improving about one day per decade in last several decades with advances in NWP model developments, application of more satellite observations, and high-performance computing power [93]. Especially after global satellite observations are applied in the NWP data assimilations, the NWP skills are proven very accurate for 3-day forecast, highly accurate for 5-day forecasts and very useful for

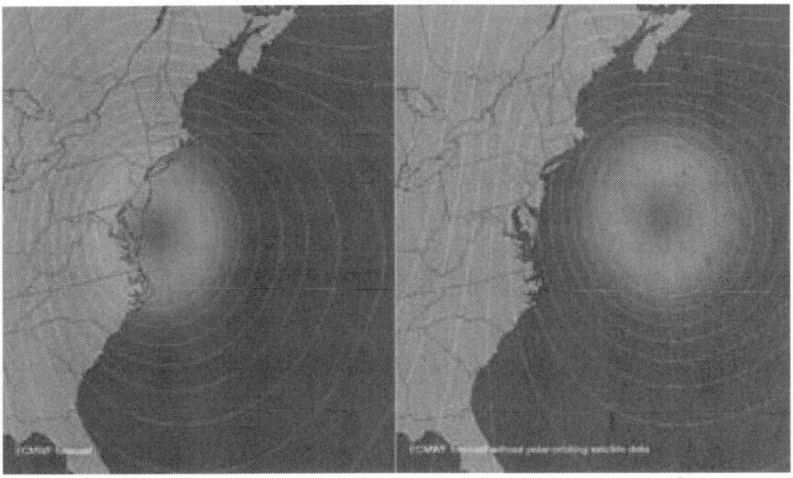

Figure 11. This image uses the model output from the ECMWF experiments, showing where Sandy was predicted to be located 5-days out with the normal satellite data inputs into the model (left) and without any polar-orbiting satellite data (right). Both position and intensity forecasts were affected—Sandy stays out to sea without the polar-orbiting satellite data, and the closer isobar lines encircling the storm also imply a more organized and stronger system (adapted from NOAA at http://www.noaanews.noaa.gov/stories2012/20121211_poesandsandy.html).

7-day forecasts for both northern and southern hemispheres. Modern NWP models even show skill for extended forecast beyond 10 days and up to months. The advanced climate models could provide seasonal and longer time predictions with various confidence levels [94].

Accurate predictions of TC genesis, intensity, and track are crucial for preparation and mitigation of TC's impacts. The forecast skills of global NWP models were always superior in the northern hemisphere than the southern hemisphere until 1999 when global satellite measurements were successfully assimilated so that difference of the prediction skills between northern and southern hemispheres diminished [93, 95]. The role of satellite observations in NWP forecast skills is normally assessed by the observing system experiments (OSEs) in which denying or adding a set of satellite data is applied from or to a baseline observing system in order to show its impacts on the forecast skills [96–98]. The famous example of the impacts of satellite observations on the NWP forecast skills is the accurate prediction of Hurricane Sandy's left (westward) turn to make landfall on the New Jersey coast for 7–8 days in advance by the European Center for Medium-Range Weather Forecasts (ECMWF) [99]. The OSE analyses show that the storm's landfall would be reasonable without observations from geostationary satellites; however, the prediction would not be very useful for 4–5 days before its landfall without measurements from polar-orbital satellites assimilated into the system. **Figure 11** presents a comparison of the predicted Hurricane Sandy's positions before its landfall with and without the polar-orbital satellites. It clearly proves that the storm's intensity and position were accurate with the LEO satellite data, while its intensity would be weak and its position offshore without satellite data assimilations.

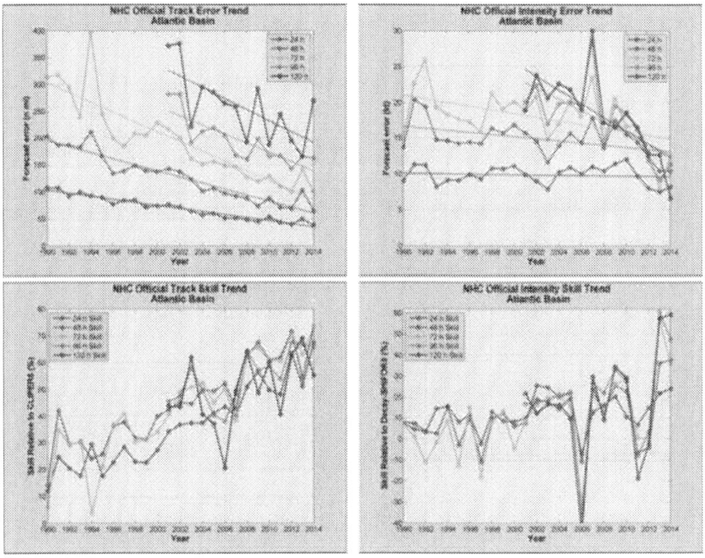

Figure 12. Error trends of the NHC official TC forecast track and intensity at 24, 48, 72, 96, and 120 h for Atlantic basin. Top-left pane is for TC track forecast error trend, while bottom-left panel is for the track forecast skill trend. The right panel is same as the left panels except for TC intensity forecast (adapted from Cangialosi and Franklin (2015)).

More accurate forecasts of TC's intensity and track will have to come from the cloud-resolving models such as the NOAA hurricane weather research and forecasting (HWRF) system and the NRL-MRY Coupled Ocean/Atmosphere Mesoscale Prediction System for Tropical Cyclones (COAMPS-TC) because of the TC's strong intensity and small spatial size. Thus, a high spatial resolution is required in making accurate TC simulations and predictions. The error trends of the TC's track forecasts have been consistently and significantly reduced in last few decades, while improvements on error trends of the TC's intensity forecasts are not so great [100]. **Figure 12** is an example of the error trends of the National Hurricane Center (NHC) TC official track and intensity forecasts over the Atlantic basin. It demonstrates that the TC track forecast errors are substantially decreased (>50%) from the 1990s to 2014, especially for 4–5 day forecasts. The track error is only around 35, 60, 80, 140, and 190 nm for 24, 48, 72, 96, and 120 hr forecasts, respectively. The track forecast skill increased from about 10% in 1990 to 70% in 2014. However, a decrease of the TC intensity forecast error is very small at 24 hr, small at 48 and 72 hr, while substantial at 96 and 120 hr. The associated TC intensity forecast skill also increases accordingly. Therefore, the TC intensity forecast skill has improved slightly; however, these improvements are still statistically significant [101, 102].

Although improvement on the TC's intensity forecast is relatively small compared with the TC track forecast, accuracy of the TC intensity forecast is obviously improved since 2001 when more satellite observations have been systematically properly applied in NWP simulations.

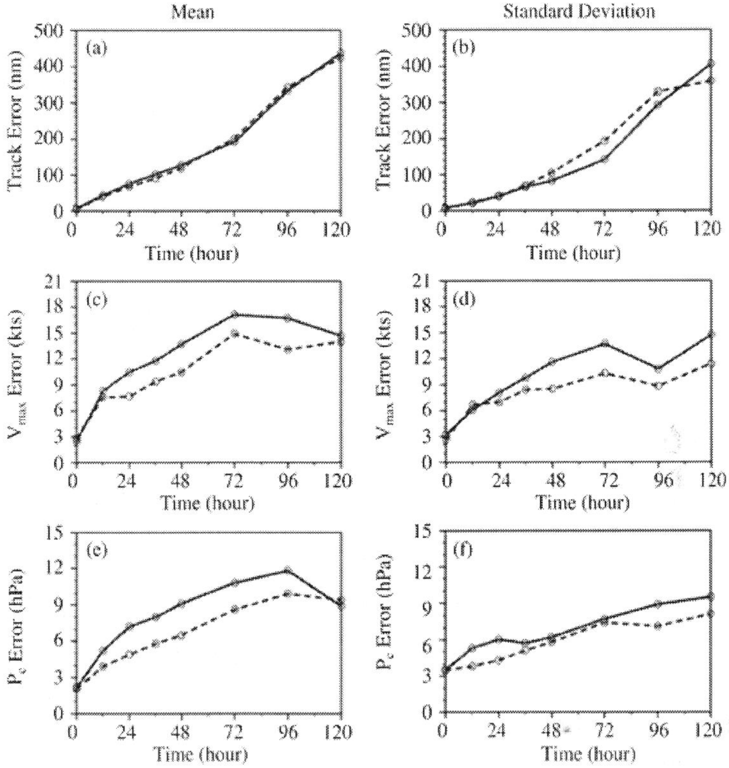

Figure 13. Mean forecast errors and standard deviations as functions of forecast lead time for TC track (a and b), maximum wind V_{max} (c and d), and minimum center pressure P_c (e and f) of CTRL2 (solid) and CTRL2 + ATMS (dashed) of the four landfall storms (adapted from Zou et al. (2013). Reproduced by permission of American Geophysical Union).

One of the important developments is to directly assimilate satellite observations into the cloud-resolving models. For example, a recent study indicates that introduction of the ATMS data into the HWRF system has significant impacts on forecasts of hurricane intensity and track [103]. Four landfall Atlantic hurricanes (Beryl, Debby, Isaac, and Sandy) in 2012 were investigated using two sets of comparisons of four experiments: CTRL1 is for assimilations of the conventional data, GPS RO data and ASCAT surface winds; CRTL1 + ATMS is for CRTL1 plus additional ATMS data; CRTL2 is for experiment setting of CRTL1 plus additional AMSU-A, AIRS and HIRS data; CRTL2+ATMS is for CRTL2 plus additional ATMS data. Results show a reduced bias of TC track for CRTL2 than CRTL1 so that impacts of the polar-orbital satellite data on TC track are further validated. The assimilation of additional ATMS data further reduced the bias of TC tracks and intensity as well as increased its lead time of forecasting. **Figure 13** presents the combined comparison results on track errors and standard deviations of the four TC forecasts between CRTL2 and CRTL2 + ATMS. The track errors are similar because of the abundant polar-satellite measurements, while the errors of maximum wind speed (V_{max}) and minimum center pressure (P_c) are significantly reduced. Although the standard deviation of track forecast is slightly large mainly due to the deteriorated Debby track forecasts, the overall V_{max} and P_c errors are obviously reduced.

The COAMPS-TC system developed at NRL-MRY has been transitioned to operations for real-time TC forecasts for several hurricane seasons at a spatial resolution of 5 km and systematically evaluated for large samples of TC forecasts over Atlantic and West Pacific basins [104]. Results demonstrate the accurate predictions of TC track and intensity, as well as the sea surface temperature cooling response to the storm, indicating the capability of the COAMPS-TC system to realistically capture characteristics of the ocean surface waves and their interactions with boundary layers above and below the ocean surface. There are more satellite measurements than what are actually assimilated into the models. Proper utilization of satellite data with positive impacts on forecast skills still requires more investigations and validations.

7. Multi-sensor-based TC monitoring and tracking

Observation from the LEO satellite PMW sensors is the best way for detection and monitoring of the global TC activities. **Figure 14** is an example of total orbits of GPM GMI observations at 89 GHz horizontal polarization on March 31, 2014. It is obvious that there are good coverages at high latitudes at daily scale, but there are large gaps in tropical and mid-latitudes. Therefore, multi-PMW sensors are required to have a reasonable coverage to monitor and track the global TC activities. Six PMW sensors could generally provide at least one measurement over a location every 3 hr. However, the high frequency channel of the PMW sensors onboard

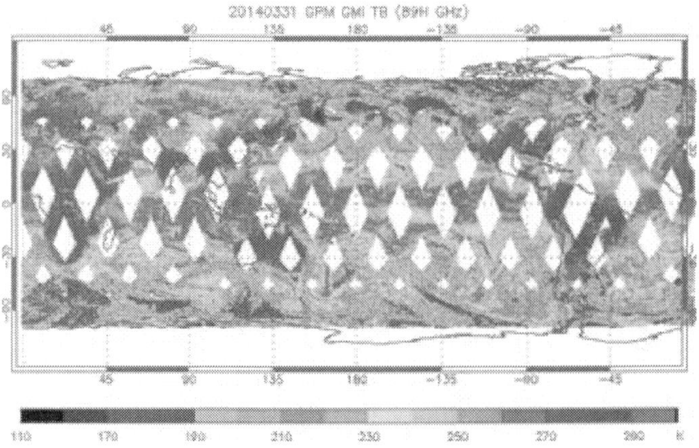

Figure 14. Example of one day satellite orbital measurements of GPM GMI 89 GHz at horizontal polarization on March 31, 2014.

different satellites is different. For example, TMI and SSM/I are at 85 GHz while SSMIS is at 91 GHz; AMSR-E, AMSR2, and GMI are at 89 GHz. These frequency differences could lead to T_B discrepancies up to 13K, which could mislead none-expert analysts in monitoring the TC's intensity, structure, and development. In order to have consistent T_Bs from different PMW sensors for improved monitoring and tracking of global TC activities, a physically based calibration scheme to calibrate T_Bs from 85 or 91 GHz TB into 89 GHz has been developed by utilizing outputs of the cloud-resolving model simulations for convective cloud systems and the associated radiative transfer model simulated T_Bs [105]. Thus, the resultant unified T_Bs at 89 GHz will be consistent among all PMW sensors.

Figure 15 presents T_B differences at horizontal polarization for TMI 85 and SSMIS 91 GHz against AMSR-E 89 GHz under four classified clouds: large rain, light rain, cloudy, and clear sky. The associated fitting curves are components of the unified calibration scheme. It is evident that the T_B differences between TMI 85 and AMSR-E 89 GHz-H at heavy rain situations could be as large as 13K. The difference is decreased to 1.5K after applying this calibration scheme. By same token, the T_B differences between SSMIS 91 GHz and AMSR-E 89 GHz are also decreased from 3K to 0.5K. The impacts of this calibration scheme are significant in improving monitoring and tracking of the global TC intensity, structure, and development because of the unified T_Bs from different PMW sensors. **Figure 16** shows an example of comparison of the observed Hurricane Igor T_Bs from TMI 85 GHz-H and AMSR-E 89 GHz-H before and after the calibration. Without the calibration (top left panel), the T_B patterns from TMI 85 GHz-H indicate a false TC intensification in four minutes because of the enhanced eyewall in red color from AMSR-E 89 GHz-H (bottom left panel). This misleading is caused by the T_B differences due to their frequency shift. With the calibration (bottom right panel), the T_B distribution patterns are very close to those observed by AMSR-E 89 GHz. The top right panel is T_B corrections due to the frequency shift. In addition, this unified calibration scheme has been applied to create a self-consistent T_B database for TCs observed by all PMW sensors, including a TC center fixing algorithm and high quality interpolation scheme. The new database can be utilized for climatology studies of TC structure, intensity, and life cycles [68, 69]

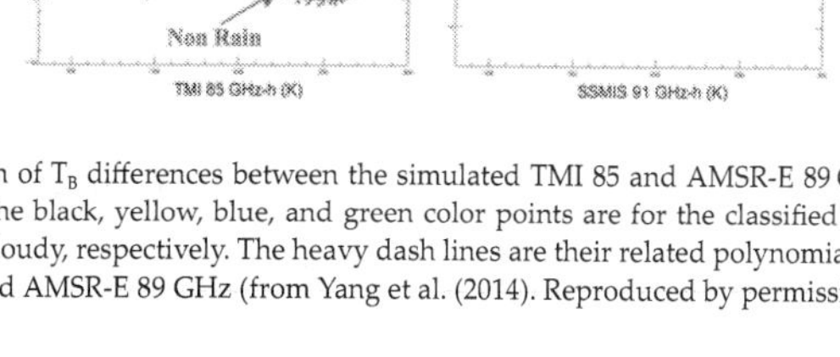

Figure 15. (a) Comparison of T_B differences between the simulated TMI 85 and AMSR-E 89 GHz H pol for Hurricane Bonnie and squall line. The black, yellow, blue, and green color points are for the classified cloud conditions of rain, light rain, non-rain, and cloudy, respectively. The heavy dash lines are their related polynomial fitting lines. (b) Same as (a) except for SSMIS 91 and AMSR-E 89 GHz (from Yang et al. (2014). Reproduced by permission of Remote Sensing).

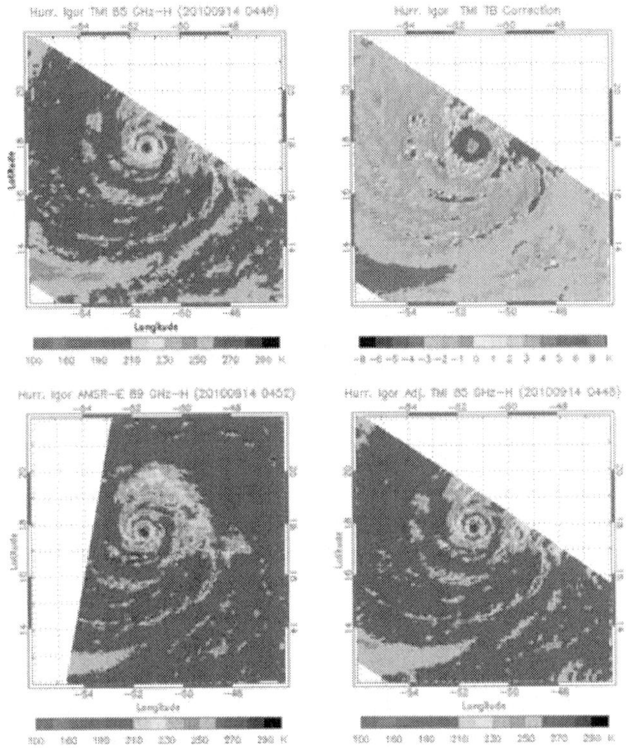

Figure 16. Impact of the newly developed physically based calibration scheme on hurricane Igor: (upper-left panel) original TMI 85 GHz-H pol; (bottom-left panel) original AMSR-E at 89 GHz-H pol; (upper-right panel) T_B correction distribution; and (bottom-right panel) calibrated TMI 89 GHz-H pol (from Yang et al. (2014). Reproduced by permission of Remote Sensing).

The GEO IR/VIS sensors are also important in monitoring the global TC activities because the LEO PMW sensors are limited. The IR/VIS sensors can fill the gaps missed by PMW sensors. The precise TC center position is the most important index not only in monitoring and tracking of TC lifecycles, but also in improving TC forecasts. The Automated Rotational Center Hurricane Eye Retrieval (ARCHER) algorithm has been developed to automate-objectively determine the TC center from 85–92 GHz channels of PMW imagers [106]. This algorithm has been applied at NRL-MRY (http://www.nrlmry.navy.mil/TC.html) and Cooperative Institute for Meteorological Satellite Studies (CIMSS) at University of Wisconsin-Madison (http://tropic.ssec.wisc.edu) for operational TC monitoring and tracking. Its updated version (ARCHER-2) is now available for including LEO 37 GHz PMW imagers, GEO IR/VIS imagery, and scatterometers [107]. It also produces a quantitative expected error estimate used for evaluation of the suitability of the estimated TC centers. The multi-satellite and multi-sensor-based TC track called the ARCHER-track is able to provide a fast access to additional TC center

http://dx.doi.org/10.5772/64114

positions for operational forecasting processes. An example of the ARCHER-track for Hurricane Michael (**Figure 17**) presents the TC centers during its lifecycle from different sensors compared with the best official TC track. It demonstrates the TC centers from PMW sensors have smaller errors than from IR/VIS sensors, although these TC centers from IR/VIS are still accurate. In addition, the IR/VIS-based TC positions provide important information to fill the gaps missed by the limited PMW sensors.

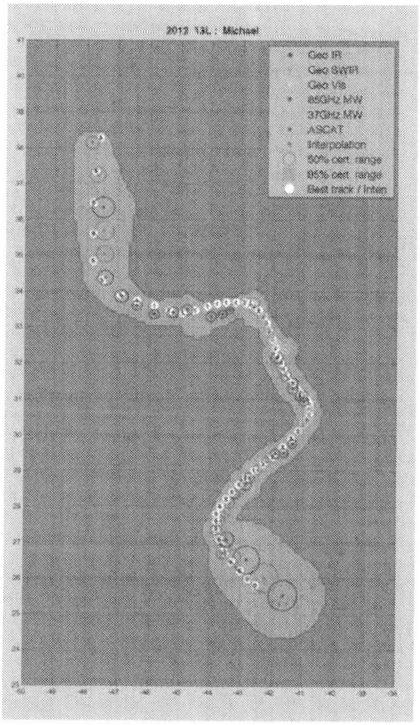

Figure 17. Example of the ARCHER-Track product for Hurricane Michael (2012). Components of the graphic are explained in the top-right legend. Inside the white circles, D = tropical depression, S = tropical storm, 1 = category-1 hurricane, etc. (from Wimmers and Velden (2016). ©American Meteorological Society. Used with permission).

Figure 18 presents another example of a multi-satellite PMW sensor-based TC imagery at 89 GHz-H for a 2014 typhoon Rammasun to display its structure, intensity, development, and tracking. ARCHER is used to find the TC centers while the multi-sensor calibration scheme is applied to generate the unified T_Bs at 89 GHz. In addition, the Backus-Gilbert interpolation scheme is utilized in providing high spatial resolution PMW T_B images [108]. The 4° × 4° boxes centered at the TC eye positions from all PMW sensors are extracted and overplayed into one image to exhibit a summary overview of the TC structure, intensity, and tracking during its lifecycle. It clearly shows evaluations of TC's key characteristics for purposes of a global TC monitoring and tracking. This new live TC tracking imagery will be added into the NRL-MRY TC web page in the near future.

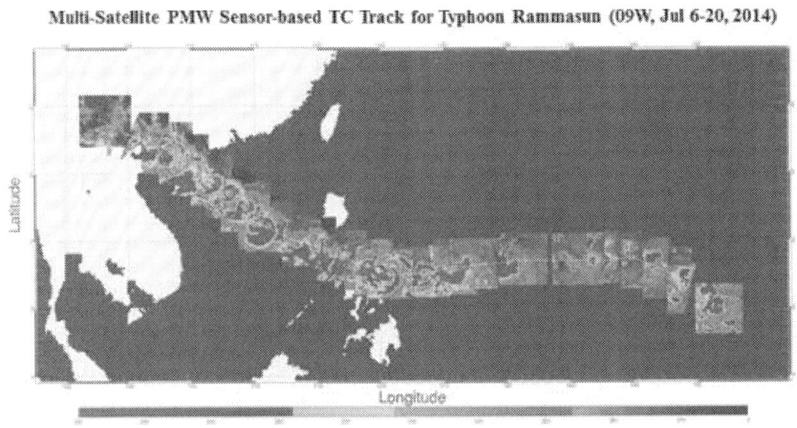

Figure 18. Example of the multi-satellite PMW sensor-based TC track for 2014 West Pacific Typhoon Rammasun (09W). The light white line is the JTWC near real-time TC track.

8. Summary

TC is one of the most destructive weather phenomena. It is initiated in tropical oceans and has a lifecycle mostly over water surface with unique horizontal characteristics of eyewall, spiral convective zones, and a vertical warm core. Satellite remote sensing is the only way to provide complete observation and monitoring of the global TC activities. The GEO IR/VIS is very useful in monitoring TC activities but not in providing accurate estimates of the TC center locations and intensity. The LEO PMW sensors are better suited for detecting TC genesis, development, and structures because of their ability to measure the atmospheric profiles. TC structure and intensity can be estimated from the PMW measurements.

Heavy precipitation is another important feature of TC activities. The abundance of TC rainfall is crucial to the drought-impacted regions because even one TC precipitation process could lead to significant relief to the severe drought situation. However, the large amount of rainfall from TC activities is also one of TC's impacts for loss of human lives and property damages. The asymmetric property of TC rainfall makes it hard to predict TC rainfall distribution. Although accurate rainfall retrievals from PMV sensors and the modern TC rainfall prediction schemes have led to reasonable TC rain forecasts, a more consistent TC rainfall from various PMW sensors and the TC diurnal characteristics are required to make further advances in TC rainfall forecasts.

Satellite remote sensing is very important in improving TC forecasts with the data assimilation process. The near real-time measurements of accurate atmospheric conditions from LEO and GEO sensors are used to improve the NWP model's initial conditions and to minimize the innovation error for better forecasts. The LEO PMW sounding sensors are especially critical in improving weather forecasts because of their ability to provide accurate atmospheric temperature and humidity profiles. The TC track forecast errors have been gradually and substantially reduced in past decades with the improved NWP models and the data assimilation schemes. Although deduction of the TC intensity forecast errors is also statistically significant, the amplitude of its improvements is much smaller than that for the TC track forecast errors. Future efforts on optimum selections of the combined satellite sensor channels which have positive impacts and better data assimilation methods are necessary in addition to improvements in the next generation NWP models and developments as well as future advanced satellite sensors onboard adequate satellites for better spatial and temporal global coverage.

Data fusion from multi-satellite sensors is the only way to provide a global coverage of TC activities. The LEO PMW sensors have advantages in high spatial resolution for TC structures, accurate TC positions, intensity analysis, and precipitation distributions, but they lack in temporal observations because each polar-orbital satellite could provide measurements only twice over a location per day. The LEO IR/VIS sensors have advantages in frequent observations of TC activities, but they lack in accurate TC eye positions, intensity analysis, and horizontal structures. ARCHER is an advanced algorithm in fixing the TC center positions from both PMW and IR/VIS sensors in near real-time with high confidence. The ARCHER track provides excellent TC positions for monitoring of TC activities and initialization in model TC data assimilation processes. The TC live track from PMW sensors will display evolutions of TC structures and intensity for purposes of better monitoring and forecasts.

Acknowledgements

The authors would like to thank the financial supports from the ONR project "tropical cyclone structure and intensity" and NRL base project "hurricane eyewall dynamics" (PE 61153N), and the NASA global precipitation measurement (GPM) project.

Author details

Song Yang* and Joshua Cossuth

*Address all correspondence to: song.yang@nrlmry.navy.mil

Naval Research Laboratory, Monterey, CA, USA

References

[1] Blake, E.S., Landsea, C.W., and Gibney, E.J.; National Hurricane Center (August 2011). The Deadliest, Costliest, and Most Intense United States Tropical Cyclones from 1851 to 2010 (And Other Frequently Requested Hurricane Facts) (PDF) (NOAA Technical Memorandum NWS NHC-6). United States National Oceanic and Atmospheric Administration's National Weather Service. Accessed on April 28, 2016 http://www.webcitation.org/6CUXzlU54

[2] Ten Years of Hurricanes and Tropical Storms in One Graphic. Accessed on April 28, 2016 http://news.nationalgeographic.com/2015/08/140829-ten-years-of-hurricanes-tropical-storms-graphic/

[3] Gray, W.M.: Hurricanes: their formation, structure, and likely role in the tropical circulation. In: Meteorology over the Tropical Oceans, D.B. Shaw, Ed., Roy. Meteor. Soc., 1979; 155–218 pp. James Glaisher House, Grenville Place, Bracknell, Berkshire, RG12 1BX.

[4] Gray, W.M.: The formation of tropical cyclones. Meteorol. Atmos. Phys., 1998; 67, 37–69.

[5] TIROS – NASA Science: Accessed on April 28, 2016 http://science.nasa.gov/missions/tiros/

[6] Sadler, J.C.: The first hurricane track determined by meteorological satellite. Preprints from the 2nd Technical Conference on Hurricanes, Miami Beach, FL, Amer. Meteor. Soc., 1961. 13 pp.

[7] GOES history: Accessed on April 28, 2016 http://www.goes-r.gov/mission/history.html

[8] Himawary-8: Accessed on April 28, 2016 http://www.data.jma.go.jp/mscweb/en/operation8/index.html

[9] Howkins, J.D., Turk, F.J., Lee, T.F., and Richardson, K.: Observations of tropical cyclones with the SSMIS. IEEE Trans. Geosci. Remote Sens., 2008; 46, 901–912.

[10] Cooperative Institute for Meteorological Satellite Studies (CIMSS) tropical cyclone web page: Accessed on April 28, 2016 http://tropic.ssec.wisc.edu/

[11] Hollinger, J.P., Peirce, J.L., and Poe, G.A.: SSM/I instrument evaluation. IEEE Trans. Geosci. Remote Sens., 1990; 28, 5, 781–790.

[12] Jezek, K.C., Merry, C., Cavalieri, D., Grace, S., Bedner, J., Wilson, D., and Lampkin, D.: Comparison between SMMR and SSM/I passive microwave data collected over the Antarctic ice sheet. Byrd Polar Research Center, The Ohio State University, Columbus, OH., BPRC Technical Report Number 91-03, ISSN: 1056-8050, 1991.

[13] Boucher, D., Poe, G., et al.: Defense meteorological satellite program special sensor microwave imager sounder (F-16) calibration/validation. Final Report, November 2005.

[14] AMSR-E: Accessed on April 28, 2016 http://wwwghcc.msfc.nasa.gov/AMSR/

[15] AMSR2: Accessed on April 28, 2016 http://suzaku.eorc.jaxa.jp/GCOM_W/w_amsr2/whats_amsr2.html

[16] NASA Precipitation Measurement Mission: Accessed on April 28, 2016 http://pmm.nasa.gov/waterfalls/science/trmm-gpm-missions

[17] Weng F., Zhou X., Wang X., Yang S., and Goldberg M.D.: Introduction to Sumi national polar-orbiting partnership advanced technology microwave sounder for numerical weather prediction and tropical cyclone applications. J. Geophys. Res., 2012. 117, D19112, doi: 10.1029/2012JD018144.

[18] Herndon, D. and Velden, C.: Estimating tropical cyclone intensity using the SSMIS and ATMS sounders. Ext. Abst. 30th AMS Hurr. Conf., Ponte Verde Beach, FL., P1.21, 5p, 2012. [Available online at https://ams.confex.com/ams/30Hurricane/webprogram/Paper205422.html]

[19] Kidder, S.Q., Goldberg, M.D., Zehr, R.M., DeMaria, M., Purdom, J.F., Velden, C.S., Grody, N.C., and Kusselson, S.J.: Satellite analysis of tropical cyclones using the Advanced Microwave Sounding Uint (AMSU). Bull. Amer. Meteor. Soc., 2000; 81, 1241–1259.

[20] Huffman, G.J., Bolvin, D.T., Braithwaite, D., Hsu, K., Joyce, R., Kidd, C., Nelkin, E.J., Xie, P.: NASA Global Precipitation Measurement (GPM) Integrated Multi-satellitE Retrievals for GPM (IMERG). Algorithm Theoretical Basis Document (ATBD) Version 4.5, 2015. Accessed on June 10, 2016 Available online at https://pmm.nasa.gov/sites/default/files/document_files/IMERG_ATBD_V4.5.pdf

[21] Rogers, R.F., Marks, F.D., and Marchok, T.: Tropical cyclone rainfall. In: Encyclopedia of Hydrological Sciences, M.G. Anderson, Ed. John Wiley and Sons, Chicester, UK, 2009. doi:10.1002/0470848944.hsa030

[22] Dvorak, V.F.: Tropical cyclone intensity analysis using satellite data. NOAA Technical Report NESDIS 1984; 11, 1–47.

[23] Velden, C., and coauthors: The Dvorak tropical cyclone intensity estimation technique: a satellite-based method that has endured for over 30 years. Bull. Amer. Meteor. Soc., 2006; 87, 1195–1210.

[24] Knapp, K.R., Kruk, M.C., Levinson, D.H., Diamond, H.J., and Neumann, C.J.: The International Best Track Archive for Climate Stewardship (IBTrACS). Bull. Amer. Meteor. Soc., 2010; 91, 363–376.

[25] Velden, C.S., Olander, T.L., and Zehr, R.M.: Development of an objective scheme to estimate tropical cyclone intensity from digital geostationary satellite infrared imagery. Wea. Forecasting, 1998; 13, 172–186.

[26] Olander, T.L. and Velden, C.S.: The advanced Dvorak technique: continued development of an objective scheme to estimate tropical cyclone intensity using geostationary infrared satellite imagery. Wea. Forecasting, 2007; 22, 287–298.

[27] Kossin, J.P., Knaff, J.A., Berger, H.I., Herndon, D.C., Cram, T.A., Velden, C.S., Murnane, R.J., and Hawkins, J.D.: Estimating hurricane wind structure in the absence of aircraft reconnaissance. Wea. Forecasting, 2007; 22, 89–101.

[28] Knaff, J.A., Longmore, S.P., and Molenar, D.A.: An objective satellite-based tropical cyclone size climatology. J. Climate, 2014; 27, 455–476.

[29] Ritchie, E.A., Valliere-Kelley, G., Piñeros, M.F., and Tyo, J.S.: Tropical cyclone intensity estimation in the North Atlantic basin using an improved deviation angle variance technique. Wea. Forecasting, 2012; 27, 1264–1277.

[30] Olander, T.L. and Velden, C.S.: Tropical cyclone convection and intensity analysis using differenced infrared and water vapor imagery. Wea. Forecasting, 2009; 24, 1558–1572.

[31] Chao, C.-C., Liu, G.-R., and Liu, C.-C.: Estimation of the upper-layer rotation and maximum wind speed of tropical cyclones via satellite imagery. J. Appl. Meteor. Climatol., 2011; 50, 750–766.

[32] Rappaport, E.N., and Coauthors: Advances and challenges at the national hurricane center. Wea. Forecasting, 2009; 24, 395–419.

[33] Chavas, D.R. and Emanuel, K.A.: A QuickSCAT climatology of tropical cyclone size. Geophys. Res. Lett., 2010; 37, L18816, doi:10.1029/2010GL044558.

[34] Chan, K.T.F. and Chan, J.C.L.: Size and strength of tropical cyclones as inferred from QuikSCAT data. Mon. Weather Rev., 2012; 140, 811–824.

[35] Kimball, S.K. and Mulekar, M.S.: A 15-year climatology of North Atlantic tropical cyclones. Part I: size parameters. J. Climate, 2004; 17, 3555–3575.

[36] Weatherford, C.L. and Gray, W.M.: Typhoon structure as revealed by aircraft reconnaissance. Part I: data analysis and climatology. Mon. Weather Rev., 1988, 116, 1032–1043.

[37] Jiang, H., Ramirez, E.M., and Cecil, D.J.: Convective and rainfall properties of tropical cyclone inner cores and rainbands from 11 years of TRMM data. Mon. Weather Rev., 2013; 141, 431–450.

[38] Li, X., Zhang, J.A., Yang, X., Pichel, W.G., DeMaria, M., Long, D., and Li, Z.: Tropical cyclone morphology from spaceborne synthetic aperture radar. Bull. Amer. Meteor. Soc., 2013; 94, 215–230.

[39] Hawkins, J.D., Lee, T.F., Turk, J., Sampson, C., Kent, J., and Richardson, K.: Real-time internet distribution of satellite products for tropical cyclone reconnaissance. Bull. Amer. Meteor. Soc., 2001; 82, 567–578.

[40] Knaff, J.A., DeMaria, M., Sampson, C.R., and Gross, J.M.: Statistical, 5-day tropical cyclone intensity forecasts derived from climatology and persistence. Wea. Forecasting, 2003a; 18, 80–92.

[41] Knaff, J.A., Cram, T.A., Schumacher, A.B., Kossin, J.P., and DeMaria, M.: Objective identification of annular hurricanes. Wea. Forecasting, 2008; 23, 17–28.

[42] Lewis, B.M. and Hawkins, H.F.: Polygonal eye walls and rainbands in hurricanes. Bull. Amer. Meteor. Soc., 1982; 63, 1294–1301.

[43] Schubert, W.H., Montgomery, M.T., Taft, R.K., Guinn, T.A., Fulton, S.R., Kossin, J.P., and Edwards, J.P.: Polygonal eyewalls, asymmetric eye contraction, and potential vorticity mixing in hurricanes. J. Atmos. Sci., 1999; 56, 1197–1223.

[44] Kieper M.E. and Jiang, H.: Predicting tropical cyclone rapid intensification using the 37 GHz ring pattern identified from passive microwave measurements. Geophys. Res. Lett., 2012; 39, L13804.

[45] Willoughby, H.E., Clos, J.A., and Shoreibah, M.G.: Concentric eye walls, secondary wind maxima, and the evolution of the hurricane vortex. J. Atmos. Sci., 1982; 39, 395–411.

[46] Jorgensen, D.F.: Mesoscale and convective-scale characteristics of mature hurricanes. Part I: general observations by research aircraft. J. Atmos. Sci., 1984; 41, 1268–1286.

[47] Hazelton, A.T. and Hart, R.E.: Hurricane eyewall slope as determined from airborne radar reflectivity data: composites and case studies. Wea. Forecasting, 2013; 28, 368–386.

[48] Rogers, R., Lorsolo, S., Reasor, P., Gamache, J., and Marks, F.: Multiscale analysis of tropical cyclone kinematic structure from airborne doppler radar composites. Mon. Weather Rev., 2012; 140, 77–99.

[49] Rogers, R., Reasor, P., and Lorsolo, S.: Airborne doppler observations of the inner-core structural differences between intensifying and steady-state tropical cyclones. Mon. Weather Rev., 2013; 141, 2970–2991.

[50] Anthes, R.A.: Tropical cyclones: their evolution, structure, and effects. Meteor. Monogr., 1982; 19, Amer. Meteor. Sco., 208 pp.

[51] Senn, H.V., Hiser, H.W, and Bourret, R.C.: Studies of hurricane spiral bands as observed on radar. National Hurricane Research Project Report No. 12, 1957; 13 pp.

[52] Senn, H.V. and Hiser, H.W.: On the origin of hurricane spiral bands. J. Meteor., 1959; 16, 419–426.

[53] Willoughby, H.E.: A possible mechanism for the formation of hurricane rainbands. J. Atmos. Sci., 1978; 35, 838–848.

[54] Kidder, S.Q., Gray, W.M., and Vonder Haar, T.H.: Estimating tropical cyclone central pressure and outer winds from satellite microwave data. Mon. Weather Rev., 1978; 106, 1458–1464.

[55] Brueske, K.F. and Velden, C.S.: Satellite-based tropical cyclone intensity estimation using the NOAA-KLM series Advanced Microwave Sounding Unit (AMSU). Mon. Weather Rev., 2003; 131, 687–697.

[56] Demuth, J.L., DeMaria, M., and Knaff, J.A.: Improvement of advanced microwave sounding unit tropical cyclone intensity and size estimation algorithms. J. Appl. Meteor. Climatol., 2006; 45, 1573–1581.

[57] Herndon, D.C., Velden, C.S., Hawkins, J., Olander, T., and Wimmers, A.: The CIMSS Satellite Consensus (SATCON) tropical cyclone intensity algorithm. 29th Conf. on Hurricanes and Tropical Meteorology, Tucson, AZ, Amer. Meteor. Soc., 2010; 4D.4. [Available online at http://ams.confex.com/ams/29Hurricanes/techprogram/paper_167959.html]

[58] Hawkins, J. and Velden, C.: Supporting meteorological field experiment missions and postmission analysis with satellite digital data and products. Bull. Amer. Meteor. Soc., 2011; 92, 1009–1022.

[59] Lee, T.F., Turk, F.J., Hawkins, J., and Richardson, K.: Interpretation of TRMM TMI images of tropical cyclones. Earth Interact., 2002; 6, 1–17.

[60] Spencer, R.W., Michael Goodman, H., and Hood, R.E.: Precipitation retrieval over land and ocean with the SSM/I: identification and characteristics of the scattering signal. J. Atmos. Oceanic Technol., 1989; 6, 254–273.

[61] Wimmers, A.J. and Velden, C.S.: MIMIC: a new approach to visualizing satellite microwave imagery of tropical cyclones. Bull. Amer. Meteor. Soc., 2007; 88, 1187–1196.

[62] Bankert, R.L., and Tag, P.M.: An automated method to estimate tropical cyclone intensity using SSM/I imagery. J. Appl. Meteor., 2002; 41, 461–472.

[63] Jones, T.A., Cecil, D., and DeMaria, M.: Passive-microwave-enhanced statistical hurricane intensity prediction scheme. Wea. Forecasting, 2006; 21, 613–635.

[64] Olander, T.L., and Velden, C.S.: Current status of the UW-CIMSS Advanced Dvorak Technique (ADT). Preprints, 30th Conference on Hurricanes and Tropical Meteorology, Ponte Vedra Beach, FL, Amer. Meteor. Soc., 2012; 7C.1/P1.19. [Available online at: https://ams.confex.com/ams/30Hurricane/webprogram/Paper204529.html]

[65] Knapp, K.R.: Hurricane satellite (HURSAT) data sets: low earth orbit infrared and microwave data. Preprints, 28th Conference on Hurricanes and Tropical Meteorology, Orlando, FL, Amer. Meteor. Soc., 2008; 4B.4.

[66] Harnos, D.J. and Nesbitt, S.W.: Convective structure in rapidly intensifying tropical cyclones as depicted by passive microwave measurements. Geophys. Res. Lett., 2011;38, L07805, doi:10.1029/2011GL047010.

[67] Cossuth, J.H., Hart, R.E., Piech, D., and Murray, D.A.: An operationally-produced climatological relationship between tropical cyclone intensity and structure. Mon. Weather Rev., 2016, in revision.

[68] Yang, S., Cossuth, J.H., Richardson, K., Surratt, M., and Bankert, R.: Tropical cyclone intensity, structure and track observed with multi-satellite sensors. Conference on Remote Sensing of the Atmosphere, Clouds, and Precipitation VI, SPIE Asia-Pacific Remote Sensing, 4–7 April 2016, New Delhi, India.

[69] Cossuth, J.H.: Exploring a comparative climatology of tropical cyclone core structures. Ph.D. Dissertation, 2014. Florida State University, Tallahassee, Florida, pp. 1–185.

[70] Rappaort, E.N.: Loss of life in the United States associated with recent Atlantic tropical cyclones. Bull. Amer. Meteor. Soc., 2000; 81, 2065–2074.

[71] Yang, S. and Nesbitt, S.W.: Statistical properties of precipitation as observed by the TRMM precipitation radar. Geophys. Res. Lett., 2014; 41, 5636–5643, doi: 10.1002/2014GL060683.

[72] Tuleya, R.E., DeMaria, M., and Kuligowski, R.J.: Evaluation of GFDL and simple statistical model rainfall forecasts for U.S. lanfalling tropical storms. Weather and Forecasting, 2007; 22, 56–70.

[73] Knight, D.B. and Davis, R.E.: Climatology of tropical cyclone rainfall in the southeastern United States. Phys. Geogr., 2007; 28, 126–147.

[74] Yang, S. and Smith, E.A.: Moisture budget analysis of TOGA-COARE area using SSM/I retrieved latent heating and large scale Q2 estimates. J. Atmos. Oceanic Technol. 1999; 16, 633–655.

[75] Elsaesser, G.S. and Kummerow, C.D.: The sensitivity of rainfall estimation to error assumption in a Bayesian passive microwave retrieval algorithm. J. Appl. Meteor. Clim., 2015; 54, 408–422.

[76] Smith, E.A., et al.: International Global Precipitation Measurement (GPM) program and mission: an overview. In: Measuring Precipitation From Space: EURAINSAT and the Future, V. Levizzani, P. Bauer, and F.J. Turk, Eds. Springer, Houten, Netherlands, 2007; pp. 611–654.

[77] Hou, A.Y., Kakar, R.K., Neeck, S., Azarbarzin. A.A., Kummerow, C.D., Kojima, M., Oki, R., Nakamura, K., and Iguchi, T.: The Global Precipitation Measurement (GPM) mission. Bull. Am. Meteorol. Soc., 2013; 95, 701–722, doi:10.1175/BAMS-D-13-00164.1.

[78] Lonfat, M., Marks, F.D., and Chen, S.: Precipitation distribution in tropical cyclones using the Tropical Rainfall Measuring Mission (TRMM) microwave imager: a global perspective. Mon. Weather Rev., 2004; 132, 1645–1660.

[79] Yu, Z., Wang, Y., and Xu, H.: Observed rainfall asymmetry in tropical cyclone making landfall over China. J. Appl. Meteor. Climatol., 2015; 54,117–136.

[80] Rogers, R., Chen, S., Tenerelli, J., and Willoughby, H.: A numerical study of the impact of vertical shear on the distribution of rainfall in Hurricane Bonnie (1998). Mon. Weather Rev., 2003; 131, 1577–1599.

[81] Marks, F., Jr.: Evolution of the structure of precipitation in Hurricane Allen (1980). Mon. Weather Rev., 1985; 113, 909–930.

[82] Kidder, S.Q., Kusselson, S.J., Knaff, J.A., Ferraro, R.R., Kuligowski, R.J., and Turk, M.: The tropical rainfall potential (TRaP) technique. Part I: description and examples. Wea. Forecasting, 2005; 20, 456–464.

[83] Kidder, S.Q., Knaff, J.A., and Kusselson, S.J.: Using AMSU data to forecast precipitation from landfalling hurricanes. Preprints, Symposium on Precipitation Extremes: Prediction, Impacts, and Responses, Albuquerque, NM, Amer. Meteor. Soc., 2001; 344–347.

[84] Kidder, S.Q., Kusselson, S.J., Knaff, J.A., and Kuligowski, R.J.: Improvements to the experimental tropical rainfall potential (TRaP) technique. Preprints, 11th Conference on Satellite Meteorology and Oceanography, Madison, WI, Amer. Meteor. Soc., 2001; 375–378.

[85] Marks, F., et al.: Development of a tropical cyclone rainfall climatology and persistence (R-CLIPER) model. Preprints, 25th conference on hurricanes and tropical meteorology, San Diego, CA, 2002; 99, 327–328.

[86] Lonfat, R., Rogers, R., Marchok, T., and Marks, F.D.: A parametric model for predicting hurricane rainfall. Mon. Weather Rev., 2007; 135, 3086–3097.

[87] Langousis, A. and Veneziano, D.: Theoretical model of rainfall in tropical cyclones for the assessment of long-term risk. J. Geophys. Res., 2009; 114, D02106, doi: 10.1029/2008JD010080.

[88] Yang, S., Olson, W., Wang, J.J., Bell, T.L., Smith, E.A., and Kummerow, C.D.: Precipitation and latent heating distributions from satellite passive microwave radiometry. Part II: evaluation of estimates using independent data. J. Appl. Meteor., 2006; 45, 721–739.

[89] Yang, S., and Smith, E.A.: Mechanisms for diurnal variability of global tropical rainfall observed from TRMM. J. Clim., 2006; 19, 5190–5226.

[90] Yang, S., and Smith, E.A.: Convective-stratiform precipitation variability at seasonal scale from eight years of TRMM observations: implications for multiple modes of diurnal variability. J. Clim., 2008; 21, 4087–4114.

[91] Yang, S., Kuo, K.-S., and Smith, E.A.: Persistent nature of secondary diurnal modes in both land and ocean precipitation. J. Clim., 2008; 21, 4115–4131.

[92] Dunion, J.P., Thorncroft, C.D., and Velden, C.S.: The tropical cyclone diurnal cycle of mature hurricanes. Mon. Weather Rev., 2014; 142, 3900–3919. doi:http://dx.doi.org/10.1175/MWR-D-13-00191.1.

[93] Bauer, P., Thorpe, A., and Brunet, G.: The quiet revolution of numerical weather prediction. Nature, 2015; 525, 47–55.

[94] Flato, G., Marotzke, J., Abiodun, B., Braconnot, P., Chou, S.C., Collins, W., Cox, P., Driouech, F., Emori, S., Eyring, V., Forest, C., Gleckler, P., Guilyardi, E., Jakob, C., Kattsov, V., Reason, C. and Rummukainen, M.: Evaluation of climate models. In: Climate Change 2013: The Physical Science Basis. Contribution of Working Group I to the Fifth Assessment Report of the Intergovernmental Panel on Climate Change, Stocker, T.F., Qin, D., Plattner, G.-K., Tignor, M., Allen, S.K., Boschung, J., Nauels, A., Xia, Y., Bex, V., and Midgley, P.M., Eds. Cambridge University Press, Cambridge, United Kingdom and New York, NY, USA, 2013.

[95] Simmons, A.J. and Hollingsworth, A.: Some aspects of the improvement in skill of numerical weather prediction. Q. J. R. Meteor. Soc., 2002. 128, 647–677.

[96] Kelly, G., Thepaut, J.-N., Buizza, R., and Cardinali, C.: The value of observations. I: Data denial experiments for the Atlantic and the Pacific. Quart. J. Roy. Meteor. Soc., 2007; 133, 1803–1815.

[97] Bauer, P., and Radnoti, G.: Study on observing System Experiments (OSEs) for the evaluation of degraded EPS/Post-EPS instrument scenarios. EUMETSAT Contract EUM/CO/07/4600000454/PS, 2009.

[98] Harnisch, F., Weissmann, M., Cardinali, C., and Wirth, M.: Experimental assimilation of DIAL water vapor observations in the ECMWF global model. Quart. J. Roy. Meteor. Soc., 2011; 137, 1532–1546.

[99] McNally, T., Bonavita, M., and Thépaut, J.: The role of satellite data in the forecasting of Hurricane Sandy. Mon. Weather Rev., 2014; 142, 634–646.

[100] Cangialosi, J.P. and Franklin, J. L.: 2014 National Hurricane Center Forecast Verification Report. NOAA/NWS/NHC, 82 pp. 2015. [Available online at http://www.nhc.noaa.gov/verification/pdfs/Verification_2014.pdf]

[101] DeMaria, M., Sampson, C.R., Knaff, J.K., and Musgrave, K.D.: Is tropical cyclone intensity guidance improving? Bull. Ameri. Meteor. Soc., 2014; 95, 387–398.

[102] DeMaria, M., Knaff, J.A. and Sampson, C. R.: Evaluation of long-term trend in tropical cyclone intensity forecasts. Meteor. Atmos Phy., 2007; 97, 19–28.

[103] Zou, X., Weng, F., Zhang, B., Lin, L., Qin, Z., and Tallapragada, V.: Impacts of assimilation of ATMS data in HWRF on track and intensity forecasts of 2012 four landfall hurricanes. J. Geophys. Res. Atmos., 2013; 118, doi:10.1002/2013JD020405.

[104] Doyle, J.D., Hodur, R.M., Chen, S., Jin, Y., Moskaitis, J.R., Wang, S., Hendricks, E.A., Jin, H., and Smith, T.A.: Tropical cyclone prediction using COAMPS-TC. Oceanography, 2014; 27(3), 104–115, doi:http://dx.doi.org/10.5670/oceanog.2014.72.

[105] Yang, S., Hawkins, J., and Richardson, K.: The improved NRL tropical cyclone monitoring system with a unified microwave brightness temperature calibration scheme. Remote Sens., 2014; 6, 4563–4581, doi:10.3390/rs6054563.

[106] Wimmers, A.J. and Velden, C.S.: Objectively determining the rotational center of tropical cyclones in passive microwave satellite imagery. J. Appl. Meteor. Climatol., 2010; 49, 2013–2034.

[107] Wimmers, A.J. and Velden, C.S.: Advancements in objective multisatellite tropical cyclone center fixing. J. Appl. Meteor. Climatol., 2016; 55, 197–212.

[108] Poe, G.A.: Optimum interpolation of imaging microwave radiance data. IEEE Trans. Geosci. Remote Sens, 1990; 28, 800–810.

[109] Willoughby, H.E.: Mature Structure and Evolution. Chapter 2, Global Perspectives on Tropical Cyclones, WMO/TD-No. 693, Report No. TCP-38, World Meteorological Organization, Geneva, Switzerland, 1995.

[110] Eliassen, A.: Slow thermally or frictionally controlled meridional circulation in a circular vortex. Astrophys. Norv., 1951; 5, 19–20.

[111] Shapiro, L.J. and Willoughby, H.E.: The response of balanced hurricanes to local sources of heat and momentum. J. Atmos. Sci., 1982; 39, 378–394.

Chapter 8

Satellite Climatology of Tropical Cyclone with Concentric Eyewalls

Yi-Ting Yang , Hung-Chi Kuo , Eric A. Hendricks and Melinda S. Peng

Additional information is available at the end of the chapter

http://dx.doi.org/10.5772/64354

Abstract

An objective method is developed to identify concentric eyewalls (CEs) for tropical cyclones (TCs) using passive microwave satellite imagery from 1997 to 2014 in the western North Pacific (WNP) and Atlantic (ATL) basin. There are 91 (33) TCs and 113 (50) cases with CE identified in the WNP (ATL). Three CE structural change types are classified as follows: a CE with the inner eyewall dissipated in an eyewall replacement cycle (ERC, 51 and 56% in the WNP and ATL), a CE with the outer eyewall dissipated first and the no eyewall replacement cycle (NRC, 27 and 29% in the WNP and ATL), and a CE structure that is maintained for an extended period (CEM, 23 and 15% in the WNP and ATL). The moat size and outer eyewall width in the WNP (ATL) basin are approximately 20–50% (15–25%) larger in the CEM cases than that in the ERC and NRC cases. Our analysis suggests that the ERC cases are more likely dominated by the internal dynamics, whereas the NRC cases are heavily influenced by the environment condition, and both the internal and environmental conditions are important in the CEM cases. A good correlation of the annual CE TC number and the Oceanic Niño index is found (0.77) in WNP basin, with most of the CE TCs occurring in the warm episodes. In contrast, the El Niño/Southern Oscillation (ENSO) may not influence on the CE formation in the ATL basin. After the CE formation, however, the unfavorable environment that is created by ENSO may reduce the TC intensity quickly during warm episode. The variabilities of structural changes in the WNP basin are larger than that in the ATL basin.

Keywords: concentric eyewall, microwave satellite, ENSO

1. Introduction

Tropical cyclones (TCs), and particularly strong TCs, are observed with a concentric eyewall (CE) structure that has an inner eyewall and an outer eyewall separated by a convective minimum region [1]. A local tangential wind maximum is associated with the outer eyewall and the most rapid increase in wind speed lies on the inside of the wind maximum [1]. The outer wind maximum thus contracts and intensifies, and then the inner eyewall weakens and eventually vanishes during eyewall replacement cycle (ERC). One of the great challenges associated with TC prediction is the large variability in structure and intensity changes, and the CE formation and the ERC is a mechanism to produce such variability [1–4]. Many theories allude to the influences of both synoptic scale environmental conditions and mesoscale processes in the CE formation. Nong and Emanuel [5] showed that the CE may form due to favorable environmental condition or external forcing and wind-induced surface heat exchange instability. Examples of internal dynamics include propagating vortex Rossby waves (VRWs) that interact with a critical radius [6, 7] and axisymmetrization during a binary vortex interaction [8, 9]. Terwey and Montgomery [10] employed idealized full physics hurricane to demonstrate the secondary eyewall form at region of sufficient low-level radial potential vorticity gradient. The result highlights the VRW energy accumulation in the critical radius with a wind-moisture feedback process at the air-sea interface. Huang et al. [11] suggested that the broadening of the radial tangential wind profile above the boundary layer (BL) in a symmetric fashion can lead to BL convergence and inflow. The progressive strengthening of the BL inflow and the unbalanced BL response may lead to secondary eyewall formation. Previous observational studies indicate that the secondary eyewall can act as a barrier to the moisture inflow to the inner eyewall (e.g., [12]).

Sitkowski et al. [4] used flight-level data to study the ERC process in the Atlantic (ATL) basin. They suggested that large variances are in the ERC time requirement, the intensity, and the change in radii because CEs are not only associated with intensity but also structural changes. Maclay et al. [13] used the low-level area-integrated kinetic energy to show that while the intensity weakens during the ERC, the integrated kinetic energy and the TC size increase. Their results suggest that CE formation and ERC are dominated by internal dynamical processes. The passive microwave data can more clearly reveal the CE structure in TCs. Using microwave data between 1997 and 2002, Hawkins and Helveston [14] suggested that CEs exist with a much higher percentage (80 and 40%) in intense TCs (maximum wind > 120 kts) than previously realized in the western North Pacific (WNP) and ATL basin. As further noted by Hawkins et al. [15], there were more CE cases with large radius in the WNP than in other basins. Hawkins and Helveston [16] provided examples of different modes of CE structure, including the ERC, triple eyewalls [17], ERCs that are repeated multiple times, ERCs that are interrupted by vertical shear and landfall, and cases where an outer eyewall forms at a large radius and remains in a CE structure for a long duration. The different CE modes appear to have profound impacts on intensity and structural forecasts. This study quantitatively examines these structural and intensity changes of CE by an objective method.

http://dx.doi.org/10.5772/64354

There have been extensive studies on TCs in different El Niño/Southern Oscillation (ENSO) phases, and no significant correlation has been found between the annual TC genesis number and ENSO over the WNP basin (e.g., [18]). The annual genesis number, however, increases over the southeastern part of the WNP and decreases over the northwestern part in the El Niño (the warm episode) and a reversed situation occurred in the La Niña (the cold episode) [19, 20]. TCs tend to recurve toward higher latitudes in the periphery of subtropical high system before landfall due to the shift of the genesis region in the warm episode. In contrast, TCs tend to move more westward in the La Niña years ([20], and references therein). The mean duration of TCs over the ocean tends to be longer during the warm episode than that in the cold episode [21]. As a result, there are more intense and long-lived typhoons in the warm episode than in the cold episode [19, 22]. In the ATL basin, ENSO inhibits the formation of TCs through the enhancement of the vertical wind shear, subsidence, and reduced relative humidity in the tropical ATL [20, 23, 24]. Teleconnection theory suggests that warm-free tropospheric temperatures that are spread eastward from the Pacific by equatorial wave dynamics can be unfavorable to convection and can influence sea surface temperature (SST) in the ATL ([25], and references therein). As stated above, shifts in TC tracks and environmental conditions have been linked to phase changes in ENSO. Since the variations in the environment have been linked to CE characteristics [26], we examine frequency and storm structures of CE TCs in relation to ENSO.

In this chapter, we present the data, the objective CE identification method, and the CE structural and intensity changes in the WNP and ATL basins. The relationship between CE TCs and ENSO is discussed and a conclusion is presented at the end of the chapter.

2. Data and methodology

The passive Special Sensor Microwave/Imager (SSM/I) 85 GHz horizontal polarized orbital imagery and Tropical Rainfall Measuring Mission (TRMM) Microwave Imager (TMI) data from the polar-orbiting TRMM satellite [27] are used in this study. These data were obtained from the website of Naval Research Laboratory (NRL) Marine Meteorology Division [15, 28]. The microwave satellite images are available and online since 1997, and the TRMM satellite ended collecting data from April 15, 2015. We use the microwave satellite data to examine the characteristics of TCs with CEs in the WNP and ATL basins between 1997 and 2014 for consistent data. The National Centers for Environmental Prediction (NCEP) warm and cold episode data are based on a threshold of ±0.5°C for the Oceanic Niño Index (ONI) in the Niño 3.4 region (5°N-5°S, 120°170°W). The ONI data are a product of three-month time running mean of SST in the Niño 3.4 region and they are used to classify the environmental condition of CE formation. The Niño 3.4 SST anomaly is used because it is better correlated with overall tropical storm activity [21].

The microwave satellite images are reprocessed using the Backus-Gilbert theory of reference [29] to create high-resolution (1–2 km) products that can assist in defining inner storm structural details [14, 15]. These images are stored as 800 × 800 pixel color jpeg files that are composed with red (R), green (G), and blue (B) colors. The pixels of R, G, and B components are converted into the high-resolution brightness temperature (T_B) based on the color table in the picture. To identify CE typhoons, the T_B dataset is transformed from Cartesian to polar coordinates with the TC center as the origin. The TC center is determined based on the Joint Typhoon Warning Center (JTWC) best track data at the time closest to satellite observation as the TC center. To further smooth the data and also to be consistent with the center position uncertainty of 10 km, we employ 5 pixel averages in the radial direction and a 45-degree sector average of T_B to obtain eight radial profiles for the bin data. In each bin of the radial profile, the T_B mean value and the standard deviation (σ) within each 45-degree sector were calculated. An objective method is developed to identify the CE structure from the eight radial profiles. The method involves the following five sequential steps:

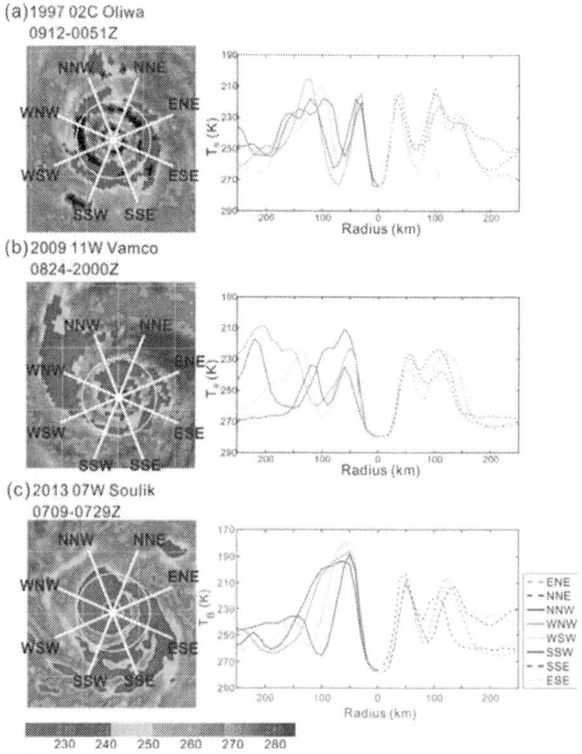

Figure 1. Color-enhanced microwave CE imageries of typhoons (a) Oliwa (1997), (b) Vamco (2009), and (c) Soulik (2013). The averaged T_B profiles of eight radial directions for Typhoon Oliwa are conformed to the CE-determined criteria. The secondary T_B minimum for Typhoon Vamco only identified spiral outer rainband. One-half symmetry of Typhoon Soulik identified CE structure (solid green: WNW; solid yellow: WSW; solid red: SSW; solid blue: NNW; dash green: ENE; dash yellow: ESE; dash red: SSE; and dash blue: NNE). Figures (a) and (b) from Yang et al. [26], courtesy of American Meteorological Society.

http://dx.doi.org/10.5772/64354

1. Within 150 km distance of the TC center, check for the existence of one local maximum T_B between two minimum T_B in each profile.

2. Check in each profile that the local T_B maximum and minimum satisfy the criteria of $T_{Bmax} \geqq \sigma_{outer\_min} + T_{Bouter\_min}$ and $T_{Bmax} \geqq \sigma_{inner\_min} + T_{Binner\_min}$.

3. For the profiles that satisfy criteria (1) and (2), check if $T_{Bouter\_min} \leqq 230K$.

4. Check if at least five out of eight sectors satisfy the above three criteria.

5. Check if the radial distance between any sectors of the two outer eyewalls is smaller than 50 km.

	Satisfy criteria (1)–(5)	**Satisfy criteria (1)–(5) but criterion (3): T_{Bouter_min}**		**Satisfy criteria (1)–(5) but criterion (4):**		**No criterion (5)**
		≤220K	**≤240K**	**≥6/8 sectors**	**≥4/8 sectors**	
WNP	113	89	130	76	145	124
ATL	50	37	60	33	64	58

Table 1. The numbers of CE cases when we use criterion (3) by making outer eyewall convection criterion 10 K weaker and stronger (240 and 220 K) than 230 K, use criterion (4) by making 4/8 and 6/8 symmetry to identified CE structure, and do not use criterion (5).

Criterion (1) identifies the existence of the structure that resembles the moat and the double eyewall in each of the eight radial profiles. Criterion (2) ensures that the moat is significant and criterion (3) ensures that the outer eyewall has strong convection. Criterion (4) ensures axisymmetry of double eyewall structure and criterion (5) ensures that the outer eyewall identified is not a spiral band. **Figure 1** provides an example of the CE TC and the no-CE TC and their associated T_B radial profiles. Three examples that have the T_B profiles of two local minima (double eyewalls) and one maximum in between the minima (the moat) are presented in **Figure 1**. The no-CE Typhoon Vamco (2009) is not classified as a CE typhoon based on our criterion (5), with the convection in the outer eyewall identified as a spiral band. Compare with CE Typhoon Oliwa (1997), Typhoon Soulik (2013) is not classified as a CE typhoon at this time based on criteria (3) and (4). If we relax criterion (3) from 230 to 240 K, or relax criterion (4) from five to four out of eight sectors, it can be considered a CE typhoon. The objective method allows us to systematically identify CE typhoons from dataset. We examined 29,785 (19,001) SSM/I and TMI satellite images in the WNP (ATL) basin from the NRL website. Out of these, 113 (50) CE cases were identified, including 17 (11) cases of multiple CE formation. There are 91 (33) CE typhoons identified in the WNP (ATL) basin. **Table 1** shows the numbers of CE cases with sensitivities in criterion (3) by making outer eyewall convection 10 K weaker and stronger (240 and 220 K) than 230 K, in criterion (4) by making 4/8 and 6/8 symmetry, and no criterion (5). Consistent with the subjective work of Kuo et al. [3], the five criteria of reference [26] ensure that the CE typhoons identified are axisymmetric with a significant moat and a strong outer eyewall while retaining enough cases for statistics. The inner eyewall radius was determined as the distance between the typhoon center to the point where $T_B = 0.5 \times \sigma_{inner} + T_{Binner}$. The moat width was determined by the distance between the points of $T_B \geqq 0.5 \times \sigma_{outer} + T_{Bouter}$ and $T_B \geqq 0.5 \times \sigma_{inner} + T_{Binner}$. Finally, the outer eyewall width was determined by the distance of the region that satisfies $T_B < 0.5 \times \sigma_{outer} + T_{Bouter}$ in the outer eyewall region. The inner eyewall radius, the moat, and the outer eyewall width were calculated by averaging the radial profiles of the eight sections as shown in **Figure 1**.

3. Structural and intensity change of concentric eyewall

In order to study the structural and intensity changes of CE TCs, we excluded the case when the TC's outer eyewall was within 200 km from land in the period of 24 h before and after CE formation, or where the satellite temporal resolution was greater than 12 h in the WNP (ATL) basin. There were 83 and 34 CE cases analyzed in the WNP and ATL basins, respectively. Three different structural change processes were defined after CE formation. The eyewall replacement cycle (ERC) cases were classified based upon the dissipation of the inner eyewall in less than 20 h after CE formation. The cases in which part of the outer eyewall dissipates within 20 h are classified as no replacement cycle (NRC) cases. The cases where the CE structure is maintained for more than 20 h are classified as concentric eyewall maintained (CEM) cases. The similar inner core size requirement was used to avoid assigning a CE TC with multiple ERC processes into one single CEM case.

The ERC classification had 42 of 83 cases and 19 of 34 cases (51 and 56%) in the WNP and ATL basins, respectively. The CEM classification had 19 and 5 cases (23 and 15%) in the WNP and ATL basins, respectively. The NRC classification had 22 and 10 cases (27 and 29%). Examples of the three classifications for the CE processes are shown in **Figure 2**. The NRC cases resemble "the shear stop ERC mode" and the CEM cases "the large radius outer eyewall and CE structure maintained for a time cases" as discussed in the study of Hawkins and Helveston [16].

Figure 3a and **b** shows the composite time series of intensities for the ERC, CEM, NRC cases as well as the average of the total CE sample. In the WNP basin, the average intensity of CEM cases is stronger than that of the ERC and NRC cases before and after CE formation. In particular, the CEM storms intensified continuously for 18 h after CE formation and maintained the intensity for another 24 h. The composite intensity of ERC and NRC cases is similar to that before CE formation. In the ATL basin, although the composite intensity of ERC is stronger than that of CEM continuously for 18 h before CE formation and 36 h after CE formation, the CEM cases maintained similar intensity before CE formation. Furthermore, the intensity of NRC decreases quickly after CE formation in both basins. **Figure 3a** and **b** indicates that a key feature of CE formation appears to be the maintenance of a relatively high intensity for a longer duration rather than a rapid intensification process to a high intensity. The stronger core intensity may play a pivotal role in the axisymmetrization dynamics of asymmetric convection outside the core to produce the CE structure [8, 9]. This asymmetric convection is also shown in **Figure 2**.

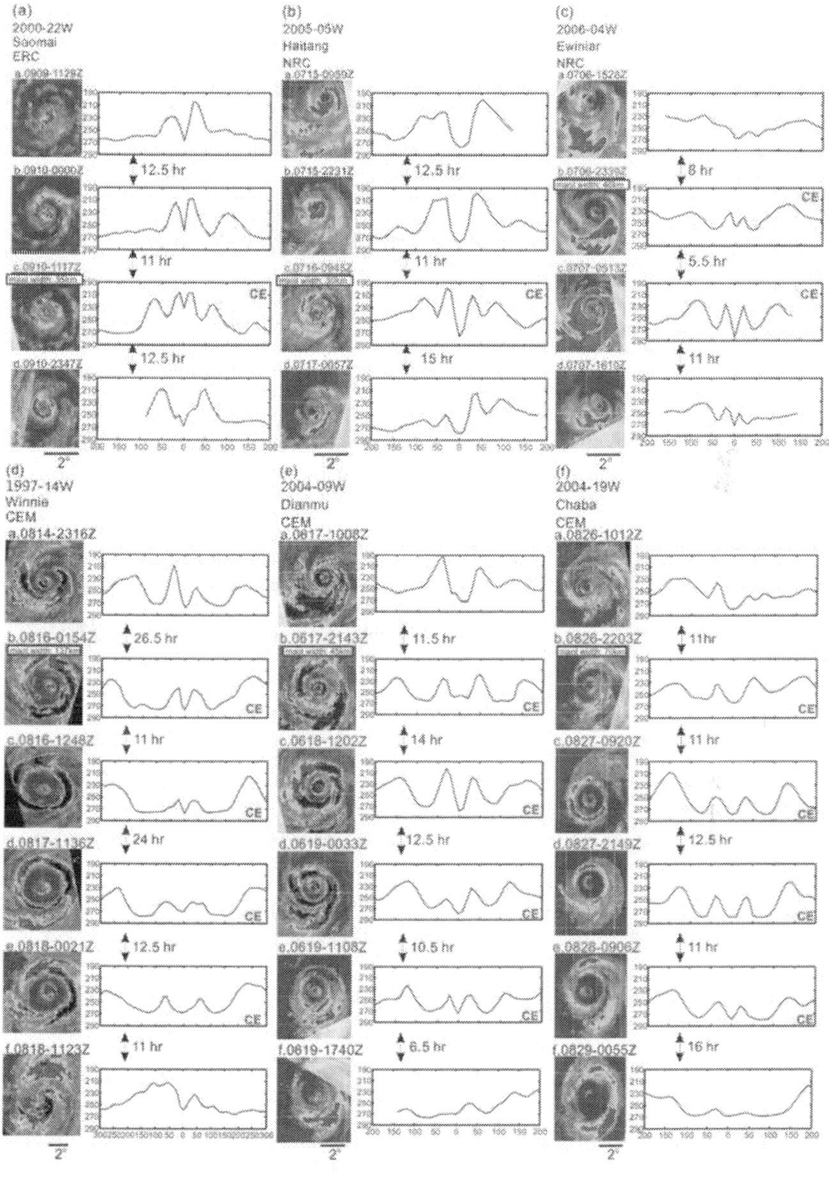

Figure 2. The imagery sequences and averaged TB radial profile for (a) Typhoon Saomai 2000-ERC, (b) Haitang 2005-NRC, (c) Ewiniar 2006-NRC, (d) Winnie 1997-CEM, (e) Dianmu 2004-CEM, and (f) Chaba 2004-CEM. From Yang et al. [26], courtesy of American Meteorological Society.

Maclay et al. [13] use aircraft data to construct the $K\text{-}V_{max}$ diagram for the intensity and structural changes. However, there is few active aircraft reconnaissance program for the WNP basin. Therefore, the structural and intensity variability is illustrated here using the $T\text{-}V_{max}$ diagram (where T is the T_B and V_{max} is the best track estimated intensity). The convective activity (CA) is indicated by the areal averaged T_B contrast to the background T_B in the 400 km square area of satellite imagery centered at the eye ($CA \equiv -\overline{T_{B1} - T_{B0}}$). The background T_{B0} is calculated as the highest 5% of T_B in the 400 km square area. The 400 km square box in general is sufficient to cover the structure of CE TCs.[1] Yang et al. [30] also used $T\text{-}V_{max}$ diagram to analyze Typhoon Soulik (2013), which had two long-lived CE episodes. **Figure 3c** and **d** shows the $T\text{-}V_{max}$ diagrams for average values of intensity and CA for the no-CE TCs with intensity category 4 or above and far from land (NCE), and CE TCs. The CE cases have stronger averaged CA, in particular, the CEM cases indicates significant CA increase 24 h after CE formation in the WNP basin. The maintenance or a slight increase of the CA for three types and a slower decrease than that of NCE cases in both basins are in general agreement with the notion that the CE TCs can lead to storm growth [13]. The decrease of areal averaged T_B and the increase of kinetic energy both occurred after the ERC process.

Figure 3e and **f** indicates that the outer eyewall width is larger with a larger moat width (R^2 = 0.5) in both basins. All the CEM cases have moat widths greater than 30 km in both basins. In particular, the CEM cases on average have slightly higher intensities, larger moat widths, and larger outer eyewall widths than those of ERC and NRC cases. The CEM cases in the ATL basin also have similar characteristics except the average intensity slightly lower than that of ERC. The ATL basin is smaller than WNP basin, only five CEM cases are classified in the ATL basin. If we choose 15 h (10 h) for CEM criteria, eight (18) cases are classified into CEM cases in the ATL basin. These CEM cases on average have higher intensity, larger moat widths, and larger outer eyewall widths. In general, the moat size and outer eyewall width are approximately 20–50% (15–25%) larger in the CEM cases than that in the ERC and NRC cases. The very large moat and outer eyewall width in the CEM cases may have some implications for the long duration of CE structure. Willoughby [31] presented a scale analysis on the validity of the balance model and the transverse circulation equation in the TC. Rozoff et al. [32] used the balanced model transverse circulation equation to study the ERC dynamics. In this manner, the balance dynamics of the CE is scale free, namely, the dynamics may occur in different scales where the balance equation assumption is valid. Thus, it is possible that the larger CE storms simply end up taking much longer time to contract due to their larger scale. Rozoff et al. [32] showed that the decay of the inner eyewall may be related to the fact that the upper warm core has a larger stabilization effect on the convection in the inner eyewall than it does on the convection in the outer eyewall. The stabilization effect of upper warm core argument cannot explain why the inner eyewall is maintained for such a long time in the CEM cases. We also note that the CE variabilities of intensity and structural changes in the WNP basin are larger than that in the ATL basin as shown in **Figure 3**.

1 Typhoons Winnie (1997) and Amber (1997) were very large, and these quantities are calculated using a 600 km square box.

http://dx.doi.org/10.5772/64354

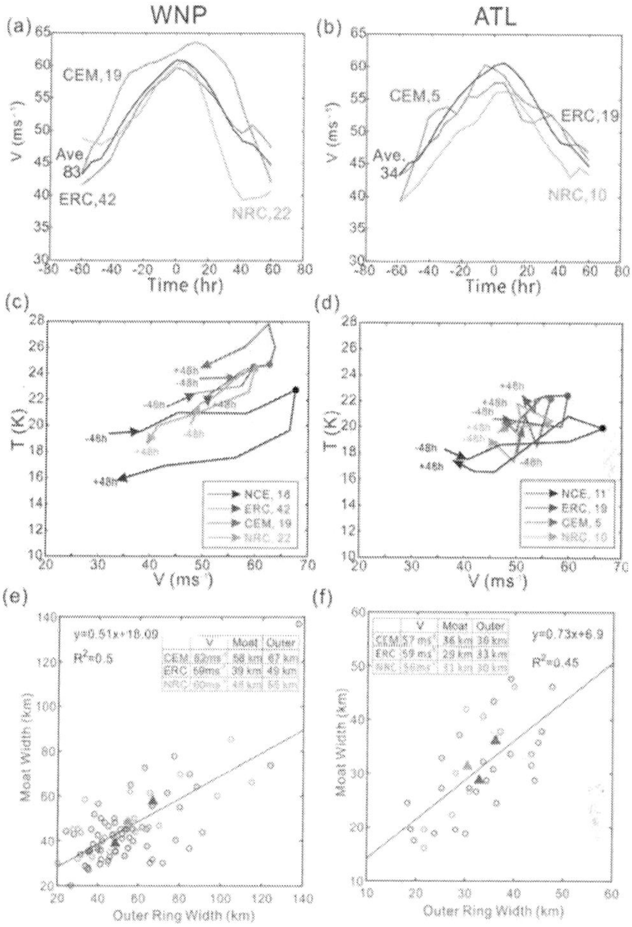

Figure 3. (a and b) Composite time series of intensity for the ERC, CEM, NRC cases, and the all CE cases in WNP and ATL basins. (c and d) The averaged T_B and intensity changes in 48 h before and after the peak intensity/CE formation for the NCE, ERC, CEM, and NRC cases in WNP and ATL basins. Case numbers are given. The dots are CE formation time for CE cases and V_{max} time for NCE cases. (e and f) Scatter diagrams of the moat width versus outer eyewall width in the WNP and ATL basins. The linear fitting line and formula are also shown. The table indicates the average intensity, moat width, and outer eyewall width of CEM, ERC, and NRC cases, respectively.

On the other hand, the occurrence of barotropic instability will invalidate the axisymmetric balance assumption. Kossin et al. [33] identified two types of barotropic instabilities in the vorticity field with CE structure: the instabilities across the outer eyewall (Type I) and across the moat (Type II) due to the sign reversal of the radial vorticity gradient. These instabilities may work against the maintenance of the CE structure. The large moat size in the CEM cases has two dynamic implications. It reduces the growth rate of the Type II instability across the moat which is favorable for the CE structure maintenance; and it also lessens the stabilization of the core vortex on the Type I instability across the outer eyewall which is not favorable for the CE maintenance. As demonstrated by Kossin et al. [33], the thicker outer eyewall is more stable for the type I instability, which is favorable for maintaining the outer eyewall structure. These observations of the large outer eyewall and moat widths are in general agreement with the concept that barotropic dynamics may play a significant role in maintaining the CE structure for CEM cases.

Finally, we note that the large moat size in the CEM cases may have an impact on the convection and subsidence in both eyewalls. The interference between the convection/subsidence couplet of the inner and outer eyewalls may be reduced when the moat size is very large. The large moat size may assist the inner core by suppressing potentially competing convection while the subsidence concentrated radial outward may make it less likely to penetrate to the eyewall. Zhou and Wang [34], in the modeling study, revealed that the demise of the inner eyewall is primarily due to the interception of the BL inflow supply of entropy by the outer eyewall. The interception process becomes inefficient when the moat size is large. **Figure 3e** and **f** suggests that the internal structure of CE TCs, such as the general high intensity with the large widths of the moat and outer eyewall, may be important for the maintenance of the CE structure in the CEM cases.

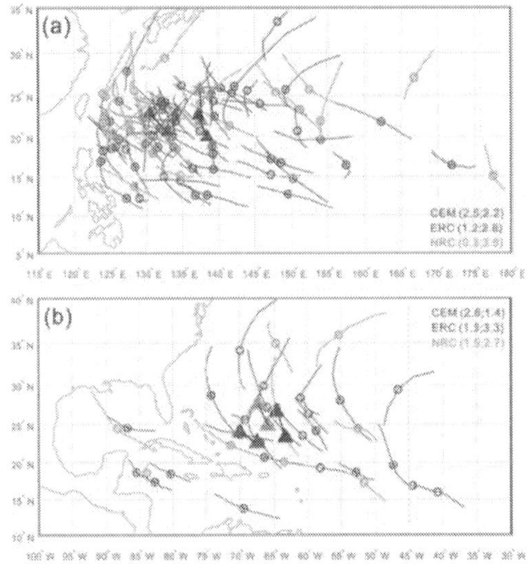

Figure 4. (a) Tracks within 48 h centered at CE formation in the WNP. The ERC, NRC, and CEM cases are represented by blue, green, and red colors, respectively. The circles with (without) a dot are the location of the secondary eyewall formation with intensity greater than or equal to (less than) category 4 on the Saffir-Simpson scale. The triangle symbols represent the composite location of CE formation and 24 h after CE formation. The average translation speed of zonal and meridional between CE formation and 24 h after CE formation is shown. (b) As in (a) but the cases in the ATL basin.

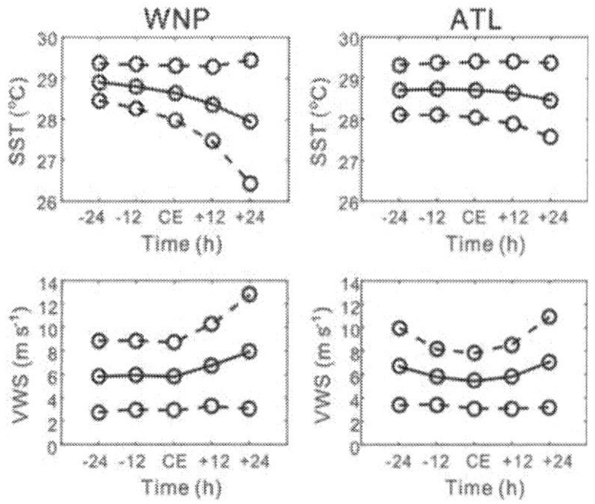

Figure 5. The time series of SST and 850–200 mb vertical shear (VWS) in the WNP and ATL basins. The solid lines mean average value. The dash lines mean average value ± 1 standard deviation.

Figure 4a shows the locations of CE formation on the tracks 24 h before and after the formation. Most CEM cases are located west of 140°E with the smallest northward translation speed of 2.2 m s^{-1} in the WNP basin. This suggests TCs tend to be more intense in the western part of Pacific after a long journey over ocean. In the ATL basin, the average location of CEM cases is farther west than that of ERC and NRC cases with the smallest northward translation speed of 1.4 m s^{-1}. Compared with ERC and NRC cases, CEM cases encounter warmer SST after CE formation. On the other hand, the NRC cases have a larger northward translation speed (3.8 m s^{-1}) in the WNP basin, and the average location of CE formation is farther north (25°N) than that of ERC and CEM cases in the ATL basin. The composite time series with respect to CE formation time for SST, relative humidity (RH), ocean heat content (OHC), and maximum potential intensity (MPI) decrease and vertical shear increases with time because TCs in general move toward the northwest direction by the Statistical Typhoon/Hurricane Intensity Prediction Scheme model data (STIPS/SHIPS; [35, 36]). The large northern translation speed of the NRC cases causes them to experience colder SST, larger vertical wind shear, smaller RH, smaller OHC, and smaller MPI 24 h after CE formation (not shown; the result of WNP cases is presented in **Figure 10** of reference [26]). These phenomena are consistent with the sharp decrease of CA and intensity in NRC cases as shown in **Figure 3**. The dissipation of the outer eyewall in the NRC cases presumably may also be related to the strong vertical shear in the high latitudes [16]. Moreover, the CEM cases were under small vertical wind shear, high SST, OHC and MPI, and high low- to mid-level RH throughout the period of CE formation. These favorable environment factors may help CEM cases maintain their intensity and eyewall structures. The environmental conditions play a role in the structural and intensity changes of CEM and NRC TCs. **Figure 5** shows that the variabilities of SST and 850–200 mb vertical wind shear during CE formation and after CE formation are larger in the WNP basin than that in the ATL basin. This results in the larger CE variabilities of intensity and structural changes in the WNP basin (**Figure 3**).

4. The relationship between CE TC and ENSO

The environmental factors and TC activities are deeply affected by ENSO and have been examined by many previous studies. We have discussed the importance of environmental factors for CE TCs in Section 3. In this section, we followed Yang et al. [38] but included CE cases in the ATL basin. Furthermore, we examined the CE TCs in different phases of ENSO. There are 46 months for five warm episodes, 62 months for four cold episodes, and 103 months for eight neutral episodes during 1997–2014 period according to NCEP data. There are 38 (18), 16 (10), and 59 (22) CE TCs occurred in the warm, cold, and neutral episodes in the WNP (ATL) basin.

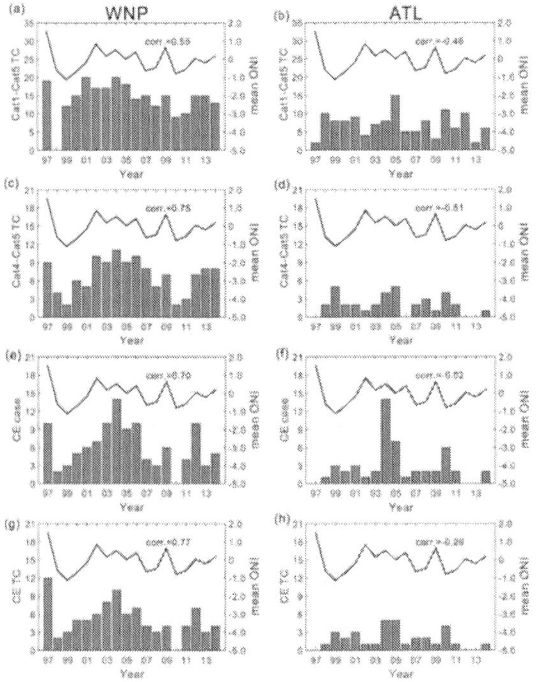

Figure 6. Number of (a) category 1–5 TCs, (c) category 4–5 TCs, (e) CE cases (multiple CE formations are included), and (f) CE TCs by years (histograms and left ordinate) and the mean ONI by year (line and right ordinate) in the WNP basin. The correlation between number and mean ONI is shown. (b, d, f, and h) As in (a, c, e, and h) but in the ATL basin.

In the WNP basin, the correlation between annual CE TCs number (CE cases) and ONI by year is 0.77 (0.70) as shown in **Figure 6**. **Figure 6** also shows that the correlation between ONI and annual strong TCs of Category 4 or stronger is 0.75, which means more intense TCs occur in the El Niño phase than that in the La Niña phase. This is in general agreement with previous studies [19, 22]. All the CE-related correlations are higher than the correlation of TC number and ONI by year (0.55). The better correlation of CE TCs and ONI may be due to the fact that the CE structure is likely to occur in strong TCs [3]. In the ATL basin, the negative correlation of annual TCs (strong TCs) and ONI is −0.46 (−0.51), which is consistent with previous studies [20, 23, 24, 37] that suggest unfavorable environment for TC formation in the El Niño phase. Moreover, the worse negative correlation of CE TCs and ONI may be because of only 33 CE TCs in the ATL basin. In addition, the better correlation with CE TCs than with CE cases may be due to the fact that multiple CE formation may be controlled by both internal and environmental factors in both basins.

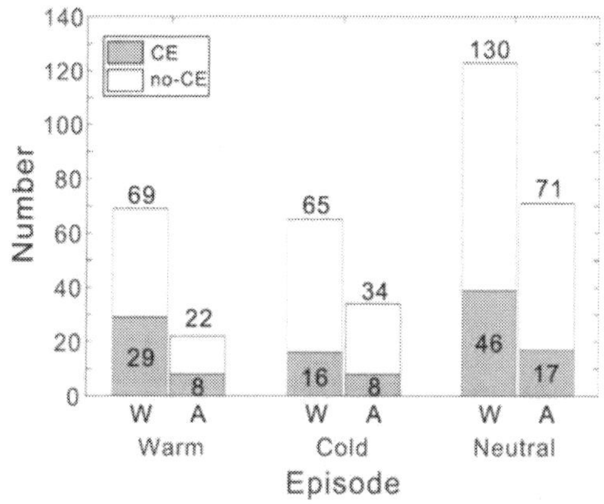

Figure 7. Histogram of the no-CE TCs and CE TCs in three different episodes. The numbers of total TC and CE TCs in each episode are indicated. The letters W and A mean WNP and ATL basin, respectively.

Figure 7 indicates that 42% (35%) of TCs possessed CE structures in their lifetimes during warm (neutral) episodes in the WNP basin. In contrast, only 25% of TCs formed CE structure in the cold episodes. In the ATL basin, 36% (approximately 24%) of TCs possessed CE structures in their lifetimes during warm (neutral and cold) episodes. Moreover, **Table 2** shows that there are 1.5 (0.5), 1.3 (0.7), and 1.0 (0.5) TC formations per month in the WNP (ATL) basin during warm, neutral, and cold episodes, respectively. The monthly CE formation frequencies are 0.8 (0.4), 0.6 (0.2), and 0.3 (0.2) in the WNP (ATL) basin during warm, neutral, and cold episodes, respectively. **Figures 6** and **7** and **Table 2** suggest that ENSO may create a better environment for CE formation in the WNP basin. In the ATL basin, the slightly higher monthly CE formation frequency during warm episode may be due to the farther south CE location as shown in **Figure 8**. For the WNP basin, **Figure 8a** suggests that CE cases during cold (warm) episodes tend to occur farther west (east). The eastward shift of the genesis region may be due to the warm sea water and moist air extending farther east (west) over the WNP during warm (cold) episodes, and the weak vertical wind shear in the southeast part of WNP during warm episodes. This result is consistent with the eastward shift of the TC genesis region in the warm episode [19, 20]. For the ATL basin, **Figure 8b** shows that the average location of CE formation during warm and cold episodes is similar. After 24 h of CE formation, however, the 850–200mb vertical wind shear in the cold episode is 2–6 m s^{-1} weaker than that in other episodes (not shown) and may help in CE maintenance. The farther south CE location during warm episode may lead to the slight CE formation frequency.

	Warm	Cold	Neutral
WNP TC number/month	69/46 = 1.5	65/62 = 1.0	130/103 = 1.3
WNP CE cases/month	38/46 = 0.8	16/62 = 0.3	59/103 = 0.6
ATL TC number/month	22/46 = 0.5	34/62 = 0.5	71/103 = 0.7
ATL CE cases/month	18/46 = 0.4	10/62 = 0.2	22/103 = 0.2

Table 2. The number of TCs, CE cases, and multiple CE cases per month during warm, cold, and neutral episodes.

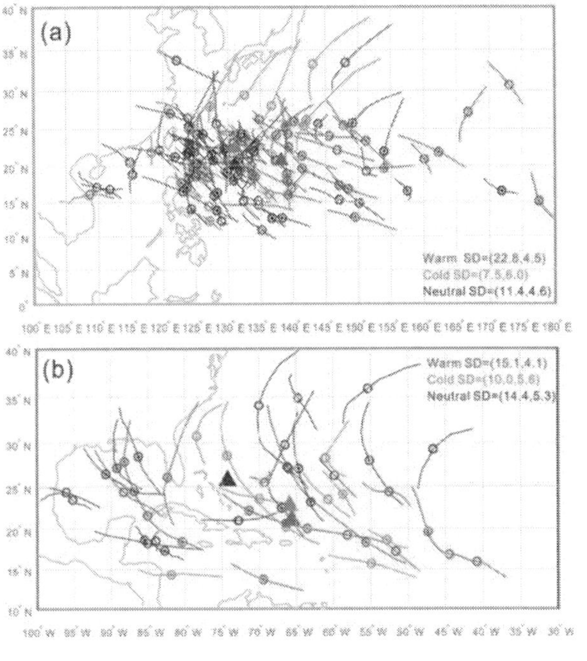

Figure 8. Same as **Figure 4** but during warm (red), cold (blue), and neutral (black) episodes. The triangle symbols represent the composite location of CE formation. The standard deviations of zonal and meridional locations are shown.

Figure 9 suggests that there are only three CEM cases during cold episode in the WNP basin. In addition, there are 12% (3%) of CE cases with more than 30 h long duration in the WNP (ATL) basin. The CE storms in the cold episode have a more rapid intensification rate than that of the warm episode before the CE formation (**Figure 10**). Specifically, there are 13 out of 16 storms in cold episode with intensity change which meet the rapidly intensifying criteria of $\Delta V_{max} \geq 19.5$ m s^{-1} in 24 h [39]. The storm intensity change after the CE formation during warm (cold) episodes often decreases slowly (quickly). The quick decline of intensity during cold episodes may be due to the encountering of unfavorable environmental factors such as the colder SST. **Figure 10b** indicates the similar trends of the average intensity before CE formation in the ATL basin. However, a rapid decreasing trend 24 h after CE formation is in the warm episode. In summary, the CE formation in the ATL basin may not be affected by ENSO because of the average location of CE formation during warm episode farther south over the ATL. After CE formation, the unfavorable environment which is created by ENSO may reduce the TC intensity quickly during warm episode.

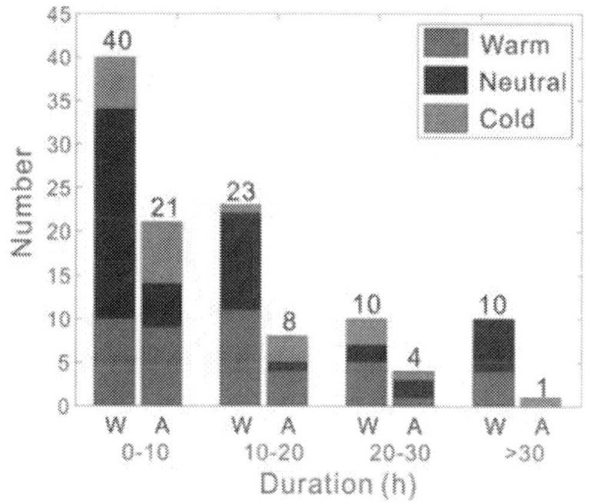

Figure 9. The number of CE cases in different episodes as a function of duration time. The letters W and A mean the WNP and ATL basins, respectively.

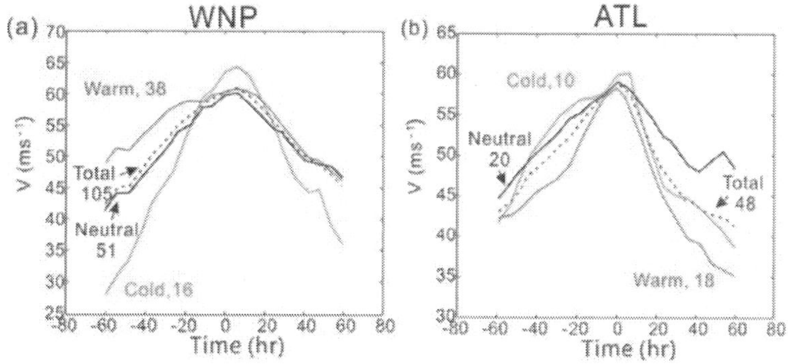

Figure 10. Same as **Figure 3** but during warm, cold, and neutral episodes and all CE cases.

5. Summary and conclusions

An objective method is developed to categorize concentric eyewall structures in the western North Pacific (WNP) TCs using the NRL SSM/I and TMI satellite imagery database. For the WNP (ATL) basin, there are 91 (33) CE TCs and 113 (50) CE cases identified from 29,785 (19,001) satellite images between 1997 and 2014. Excluding the cases that are 200 km close to landfall and cases with temporal resolution higher than12 h, 83 (34) CE cases were studied for the structural and intensity changes. The primary findings are as follows:

1. Three CE types are categorized: CE with an eyewall replacement cycle (ERC; 51 and 56% in the WNP and ATL), CE with no replacement cycle (NRC; 27 and 29% in the WNP and ATL), and CE that is maintained for an extended period of time (CEM; 23 and 15% in the WNP and ATL). The mean duration of CEM type is 32 and 27 h in the WNP and ATL basins, respectively.

2. The CEM cases have relatively high intensity, large widths of both the moat and outer eyewall that last for a long duration time. The large widths of the moat and outer eyewall may reduce the barotropic instabilities of CE storms and thus maintain the CE structures for a long time.

3. Most CEM cases in the WNP basin are located to the west of 140°E with the smallest northward translation speed of 2.2 m s^{-1}. The average location of CEM cases in the ATL basin is farther west than that of ERC and NRC cases with the smallest northward translation speed of 1.4 m s^{-1}. The NRC cases often have fast northward translation speeds and are located in higher latitudes of relative strong vertical shear and cold SST zones. The unfavorable environmental conditions thus act to weaken the convective activity for these cases. On the other hand, the CEM cases occur in favorable environment and tend to form in the warm episode of ENSO in the WNP basin. There are 12% (3%) of CE cases with more than 30 h long duration in the WNP (ATL) basin.

4. The variabilities of intensity and structural changes in the WNP basin are larger than that in the ATL basin. For example, the moat size and outer eyewall width in the WNP (ATL) basin are approximately 20–50% (15–25%) larger in the CEM cases than that in the ERC and NRC cases.

5. In the WNP basin, a very good relationship of 0.77 was found between the annual CE TC numbers and the NCEP Oceanic Niño Index (ONI). The probability of CE TCs formation is strongly influenced by the ENSO environmental factors. For example, there are 29 out of 69 (42%) TCs that possess CE structures in the warm episode. In contrast, there are only 16 out of 65 (25%) TCs that possess CE structures in the cold episode. The averaged CE formation frequencies are 0.8 and 0.3 per month in the warm and cold episodes, respectively.

6. In the ATL basin, the correlation between ONI and CE formation frequency suggests that the ENSO may not influence the CE formation. It may be due to the average location of CE formation during warm episode farther south over the ATL in the relative warm SST

http://dx.doi.org/10.5772/64354

region. After CE formation, however, the unfavorable environment that is created by ENSO may reduce the TC intensity quickly.

Acknowledgements

This work was supported by the Ministry of Science and Technology in Taiwan under grants MOST 104-2111-M-002-002-MY3 and MOST 104-2625-M-002-006, and the Naval Research Laboratory under grant N62909-15-1-2008.

Author details

Yi-Ting Yang[1], Hung-Chi Kuo[2*], Eric A. Hendricks[3] and Melinda S. Peng[4]

*Address all correspondence to: kuo@as.ntu.edu.tw

1 Taida Institute for Mathematical Sciences, National Taiwan University, Taipei, Taiwan

2 Department of Atmospheric Sciences, National Taiwan University, Taipei, Taiwan

3 Department of Meteorology, Naval Postgraduate School, Monterey, California, USA

4 Marine Meteorology Division, Naval Research Laboratory, Monterey, California, USA

References

[1] Willoughby, H. E., J. A. Clos, and M. G. Shoreibah, 1982: Concentric eye walls, secondary wind maxima, and the evolution of the hurricane vortex. J. Atmos. Sci., 39, 395–411.

[2] Black, M. L., and H. E. Willoughby, 1992: The concentric eyewall cycle of Hurricane Gilbert. Mon. Wea. Rev., 120, 947–957.

[3] Kuo, H.-C., C.-P. Chang, Y.-T.Yang, and H.-J. Jiang, 2009: Western North Pacific typhoons with concentric eyewalls. Mon. Wea. Rev., 137, 3758–3770.

[4] Sitkowski, M., J. P. Kossin, and C. M. Rozoff, 2011: Intensity and structure changes during hurricane eyewall replacement cycles. Mon. Wea. Rev., 139, 3829–3847.

[5] Nong, S., and K. A. Emanuel, 2003: A numerical study of the genesis of concentric eyewalls in hurricane. Quart. J. R. Meteor. Soc., 129, 3323–3338.

[6] Montgomery, M. T., and R. J. Kallenbach, 1997:Atheory for vortex Rossby-waves and its application to spiral bands and intensity changes in hurricane. Quart. J. R. Meteor. Soc., 123, 435–465.

[7] Peng, J., T. Li, and M. S. Peng, 2009: Formation of tropical cyclone concentric eyewalls by wave-mean flow interactions. Adv. Geosci., 10, 57–71.

[8] Kuo, H.-C., L.-Y. Lin, C.-P. Chang, and R. T. Williams, 2004: The formation of concentric vorticity structures in typhoons. J. Atmos. Sci., 61, 2722–2734.

[9] Kuo, H.-C., W. H. Schubert, C.-L. Tsai, and Y.-F. Kuo, 2008: Vortex interactions and barotropic aspects of concentric eyewall formation. Mon. Wea. Rev., 136, 5183–5198.

[10] Terwey, W. D., and M. T. Montgomery, 2008: Secondary eyewall formation in two idealized, full-physics modeled hurricanes. J. Geophys. Res., 113, D12112.

[11] Huang, Y.-H., M.-T. Montgomery, and C.-C. Wu, 2012: Concentric eyewall formation in Typhoon Sinlaku (2008). Part II: Axisymmetric dynamical processes. J. Atmos. Sci., 69, 662–674.

[12] Barnes, G. M., and M. D. Powell, 1995: Evolution of the inflow boundary layer of Hurricane Gilbert (1988). Mon. Wea. Rev., 123, 2348–2368.

[13] Maclay, K. S., M. DeMaria, and T. H. Vonder Haar, 2008: Tropical cyclone inner-core kinetic energy evolution. Mon. Wea. Rev., 136, 4882–4898.

[14] Hawkins, J. D., and M. Helveston, 2004: Tropical cyclone multiple eyewall characteristics. Preprints, 26th Conf. on Hurricane and Tropical Meteorology, Miami, FL, American Meteorological Society, P1.7.

[15] Hawkins, J. D., M. Helveston, T. F. Lee, F. J. Turk, K. Richardson, C. Sampson, J. Kent, and R. Wade, 2006: Tropical cyclone multiple eyewall characteristics. Preprints, 27th Conf. on Hurricane and Tropical Meteorology, Monterey, CA, American Meteorological Society, 6B.1.

[16] Hawkins, J. D., and M. Helveston, 2008: Tropical cyclone multiple eyewall characteristics. Preprints, 28th Conf. on Hurricanes and Tropical Meteorology, Orlando, FL, American Meteorological Society, 14B.1.

[17] McNoldy, B. D., 2004: Triple eyewall in Hurricane Juliette. Bull. Am. Meteor. Soc., 85, 1663–1666.

[18] Chen, T. C., S. Y. Wang, and M. C. Yen, 2006: Interannual variation of tropical cyclone activity over the western North Pacific. J. Climate, 19, 5709–5720.

[19] Chia, H.-H., and C. F. Ropelewski, 2002: The interannual variability in the genesis location of tropical cyclones and the northwest Pacific. J. Climate, 15, 2934–2944.

[20] Camargo, S. J., K. A. Emanuel, and A. H. Sobel, 2007: Use of a genesis potential index to diagnose ENSO effects on tropical cyclone genesis. J. Climate, 20, 4819–4834.

[21] Wang, B., and J. C. L. Chan, 2002: How strong ENSO events affect tropical storm activity over the western North Pacific. J. Climate, 15, 1643–1658.

[22] Li, R. C. Y., and W. Zhou, 2012: Changes in western Pacific tropical cyclones associated with the El Niño–Southern Oscillation cycle. J. Climate, 25, 5864–5878.

[23] Gray, W. M., 1984: Atlantic seasonal hurricane frequency. Part I: El Niño and 30-mb quasi-biennial oscillation influences. Mon. Wea. Rev., 112, 1649–1668.

[24] Chu, P.-S., 2004: ENSO and tropical cyclone activity. Hurricanes and Typhoons, Past, Present and Future, R. J. Murnane and K.-B. Liu, Eds., Columbia University Press, New York, pp. 297–332.

[25] Neelin, J. D., C. Chou, and H. Su, 2003: Tropical drought regions in global warming and El Niño teleconnections, Geophys. Res. Lett., 30 (24), 2275.

[26] Yang, Y.-T.,H.-C. Kuo, E. A. Hendricks, M. S. Peng, 2013: Structural and intensity changes of concentric eyewall typhoons in the western North Pacific basin. Mon. Wea. Rev., 141, 2632–2648.

[27] Kummerow, C., W. Barnes, T. Kozu, J. Shiue, and J. Simpson, 1998: The Tropical Rainfall Measuring Mission (TRMM) sensor package. J. Atmos. Ocean. Technol., 15, 809–817.

[28] Hawkins, J. D., T. F. Lee, F. J. Turk, C. Sampson, J. Kent, and K. Richardson, 2001: Real-time Internet distribution of satellite products for tropical cyclone reconnaissance. Bull. Am. Meteor. Soc., 82, 567–578.

[29] Poe, G., 1990: Optimum interpolation of imaging microwave radiometer data. IEEE Trans. Geosci. Remote Sens., 28, 800–810.

[30] Yang, Y.-T., E. A. Hendricks, H.-C. Kuo, and Melinda S. Peng, 2014: Long-lived concentric eyewalls in Typhoon Soulik (2013).Mon. Wea. Rev., 142, 3365–3371.

[31] Willoughby, H. E., 1979: Forced secondary circulations in hurricanes. J. Geophys. Res., 84 (C6), 3173–3183.

[32] Rozoff, C. M., W. H. Schubert, and J. P. Kossin, 2008: Some dynamical aspects of hurricane eyewall replacement cycles. Quart. J. R. Meteor. Soc., 134, 583–593.

[33] Kossin, J. P., W. H. Schubert, and M. T. Montgomery, 2000: Unstable interaction between a hurricane's primary eyewall and a secondary ring of enhanced vorticity. J. Atmos. Sci., 57, 3893–3917.

[34] Zhou, X., and B. Wang, 2011: Mechanism of concentric eyewall replacement cycles and associated intensity change. J. Atmos. Sci., 68, 972–988.

[35] DeMaria, M., and J. Kaplan, 1999: An updated Statistical Hurricane Intensity Prediction Scheme (SHIPS) for the Atlantic and eastern North Pacific basins. Wea. Forecasting, 14, 326–337.

[36] Knaff, J. A., C. R. Sampson, and M. DeMaria, 2005: An operational statistical typhoon intensity prediction scheme for the western North Pacific. Wea. Forecasting, 20, 688–699.

[37] Krishnamurthy L., G. A. Vecchi, R. Msadek, H. Murakami, A. Wittenberg, and F. Zeng, 2016: Impact of strong ENSO on regional tropical cyclone activity in a high-resolution climate model in the North Pacific and North Atlantic oceans. J. Climate, 29, 2375–2394.

[38] Yang, Y.-T.,H.-C. Kuo, E. Hendrick, Y.-C. Liu, and M. Peng, 2015: Relationship between typhoons with concentric eyewalls and ENSO in the western North Pacific basin. J. Climate, 28, 3612–3623.

[39] Hendricks, E. A., M. S. Peng, B. Fu, and T. Li, 2010: Quantifying environmental control on tropical cyclone intensity change. Mon. Wea. Rev., 138, 3243–3271.

Section 3

Tropical Cyclones, Modeling and Prediction

Chapter 9

Progress in Tropical Cyclone Predictability and Present Status in the North Indian Ocean Region

Kasturi Singh, Jagabandhu Panda,
Krishna K. Osuri and Naresh Krishna Vissa

Additional information is available at the end of the chapter

http://dx.doi.org/10.5772/64333

Abstract

Tropical cyclone (TC) is an important research area since it has a significant impact on human life, properties and environment. The researchers all over the world have been studying fundamental and advanced processes to better understand and thereby predict the genesis and evolution of TCs. This review chapter provides a brief overview on TC climatology, their basic characteristics, movement and intensification, research on structure analysis and prediction of these fascinating storms, with primary emphasis to North Indian Ocean (NIO). The role of ocean and atmosphere in determining the genesis and intensification of TCs is discussed. This chapter reviews the past and current research activities including inter-annual and intra-seasonal changes in TCs, current status of TC research using numerical weather prediction, gaps identified and relevant measures taken by the meteorological and government agencies in this direction, along with future directions in order to improve the understanding and predictability over the NIO region.

Keywords: tropical cyclone, cyclogenesis, predictability, North Indian Ocean, WRF

1. Introduction

A tropical cyclone (TC) is a cyclonic disturbance that originates over warm tropical oceans with anticlockwise (clockwise) winds around a centre of low barometric pressure in the Northern Hemisphere (Southern Hemisphere) [1]. It creates strong winds and intense precipitation in the regions around the system. There are seven global basins that conceive TCs, viz. North Atlantic Ocean, eastern and western parts of North Pacific Ocean, southwestern pacific, south-western and south-eastern Indian Ocean and North Indian Ocean (NIO) region. The cyclonic storms are often known as hurricanes and typhoons in the Atlantic and northwest Pacific, whereas TCs in other Ocean basins. The average frequency of occurrence, season and intensity of TCs vary from basin to basin. NIO basin shows bi-modal TC season with maximum frequency during post-monsoon period (October–December) and are comparatively stronger than pre-monsoon ones. Though the size of the TCs is relatively smaller and their intensity is comparatively less over NIO basin as compared to the other global basins, this region is quite important in view of the densely populated rim countries with poor socioeconomic conditions. Hence, loss of life and property is quite significant in this region. Based on the intensity, TCs formed over NIO basin can be classified into (i) depression if the associated 10-m maximum sustained wind (MSW) is in between 17 and 33 kt; (ii) cyclonic storm (CS) if MSW is in between 34 and 47 kt; (iii) severe cyclonic storm (SCS) if it has MSW of 48–63 kt; (iv) very severe cyclonic storm (VSCS) if the MSW is within the range 64–90 kt, and extremely severe cyclonic storm (ESCS) if the MSW is in range of 91–119 kt; and (v) super cyclonic storm (SuCS) if it has MSW of 120 kt or more (http://www.rsmcnew-delhi.imd.gov.in). This classification may differ from those over other global basins including that of the widely used Saffir-Simpson hurricane wind scale or SSHWS (http://www.nhc.noaa.gov/aboutsshws.php). Both types of classifications, that is, the earlier one from India Meteorological Department (IMD) and the SSHWS consider the tropical low-pressure system as a depression if MSW is <34 kt. The consideration of CS and SCS lies in the tropical storm category (MSW lies in the range 34–63 kt) of SSHWS. The VSCS category over NIO basin is similar to that of the category 1 hurricane (MSW lies in the range 64–82 kt) type. The category 3 (MSW lies in between 96 and 112 kt) and category 4 (MSW is within the range 113-136 kt) major hurricanes are comparable to ESCS over NIO basin. The IMD classification categorizes MSW above 120 kt as super cyclonic storm (SuCS), whereas SSHWS considers the desired wind speed above 137 kt for category 5 hurricane. However, the basic structure of NIO TCs is similar to hurricanes and typhoons.

Having a prolonged coast line, about 96 districts (lying within 100 km from the coast) of India are vulnerable to the occurrence of TCs with varying intensity [2]. Out of these 96 districts, ~59% are at least highly vulnerable. The number of CS and SCS with a core of MSW (between 34 and 63 kt) crossing different countries of the NIO region is found to be 504 during 1891–2015 (derived from the available IMD data at http://www.rmcchennaieatlas.tn.nic.in). Out of these, about 328 (>65%) crossed the Indian coasts, whereas 127 (>25%) have crossed the east coast of India between Gopalpur and Kolkata. In general, the proneness to TCs is quite high for the coastal districts of West Bengal, Odisha, Andhra Pradesh, and Tamil Nadu [2, 3]. In view of these, TCs over NIO basin can be considered as quite lethal and expensive natural disaster as they bring widespread destruction in these regions. The consequent loss of human life and properties impacts the economy of a country. Therefore, it is important to forecast the evolution of TCs by using numerical models as the frequency of such storms is increasing in several basins of the world in the present warming period [4]. Therefore, it is attempted to put forward the recent developments in understanding the related meteorological characteristics, their predictability, climatological aspects and the gaps identified in the area of TC research.

2. Life cycle of tropical cyclones

This section describes the general understanding about life cycle of TCs. The description includes a brief overview about their genesis, structural evolution, propagation and dissipation.

2.1. Cyclogenesis

The TC research has evolved over several decades and researchers use observations as well as numerical models for this purpose. For example, some observational studies [5–7] discuss about TC formation and evolution. The pioneering works by Gray [8, 9] have shown that the formation of TC at any location depends on six factors: (i) appropriate Coriolis parameter 'f' that is practically effective 5° away from the equator in both the hemispheres, (ii) low-level positive relative vorticity (ζ_r), that is, existence of initial disturbance, (iii) low tropospheric vertical wind shear (S_z), (iv) ocean thermal energy (E) signified by sea surface temperature (SST), that is, SST should be ≥26.5°C within vertical extent of 60 m, (v) atmospheric instability measured in terms of difference in equivalent potential temperature (θ_e) between the surface and 500 mb or $\Delta\theta_e$ and (vi) mid-tropospheric relative humidity 'RH'. Some of these aspects are discussed explicitly over the years using observations and numerical models [7, 10]. The first three parameters produce a dynamic potential ($f\zeta_r/S_z$), while the remaining three parameters yield a thermal potential ($E\ \Delta\theta_e$ RH). And, the product of dynamic and thermal potentials provides the seasonal genesis frequency.

Cyclogenesis do not occur spontaneously even if all the environmental conditions are met. Further, only about 10% of all cyclonic disturbances intensify into TCs. These low-pressure systems gradually form from a pre-existing (or initial) disturbance that consists of wind vortex and organized clouds. Thus, the necessary conditions for tropical cyclogenesis must be supported by the deep convection in the presence of a low-level absolute vorticity maximum and the initial convection must survive for sufficient time. The survival ability of the initial convection depends on 'ζ_r', 'f' atmospheric stability (defined by Brunt Vaisala frequency 'N') and depth of the system (H). This ability is defined by the Rossby radius of deformation 'L_R' typically for a large tropical cyclonic system (www.meted.ucar.edu):

$$L_R = \frac{NH}{\zeta_r + f} \tag{1}$$

The average life expectancy of a TC is about 1 week, whereas it is found that few cyclones remain active for more than 4 weeks (exact time frame may change from basin to basin) as seen in case of a hurricane, provided the system must be able to stay over the warm tropical waters. In most of the TCs, the Coriolis and centripetal forces oppose the pressure gradient force [11]. In the lowest kilometres near the surface, the frictional force destroys the gradient balance and consequently, air spirals inward towards the storm centres. The primary circulation (horizontal axisymmetric) during tropical cyclogenesis gains latent heat through the process of evaporation and exchange of sensible heat with the underlying ocean as it spirals towards the storm centre [12]. Consequently, it gains large angular momentum and kinetic energy because of the acceleration towards the low-pressure centre. The evaporation of sea spray provides the necessary moisture supply. Because of the high velocity demanded by the quasi-conservation of angular momentum, the air may not penetrate beyond some small radius. To conserve the angular momentum, the air spirals upward in the eyewall forming intense ring of cumulus cloud and a calm eye at the centre and brings in the latent heat it acquired during the upward motion in the boundary layer to the free atmosphere. Due to the cooling of this rising air, latent heat releases into the atmosphere to add more energy to the storm. Across the top of the boundary layer, the turbulent eddies generated by the mechanical mixing due to the prevailing strong winds cause a significant downward flux of sensible heat from the free atmosphere through subsidence (**Figure 1**).

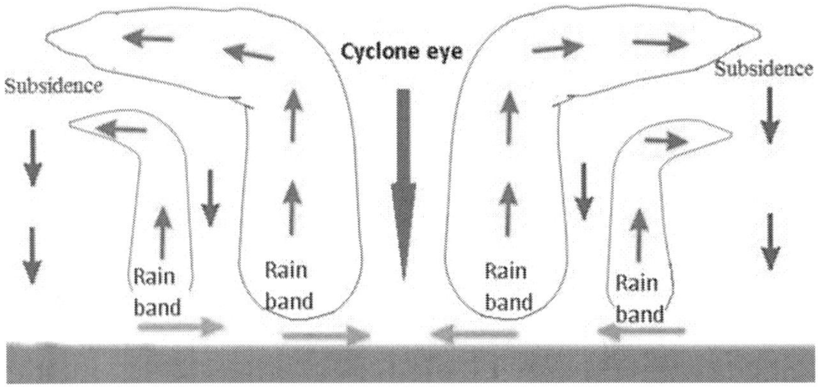

Figure 1. Vertical cross section of a mature cyclonic storm and associated basic characteristics (adopted from http://www.hko.gov.hk/informtc/nature.htm).

As the convective updrafts in the eyewall ascends to the tropopause, the latent heat is converted to sensible heat through condensation in order to provide the much-needed buoyancy for lifting air from the surface to tropopause level. After reaching at the upper level, the air turns outwards and eventually spread out at high altitudes, where it forms anticyclonic circulation and eventually the cool air above the eye begins to sink into the central core (**Figure 1**). Thus, the storm can be termed as a quasi-steady thermodynamic heat engine that is primarily driven by latent heat release. This heat engine runs between a warm heat reservoir as sea that is at ~300°K and a cold reservoir located at 15–18 km up in the troposphere having a temperature ~200°K. A baroclinic structure is maintained by the latent heat release in the warm core, which is continuously converted to kinetic energy that is responsible to drive the TC.

Apart from the basic factors discussed so far, there is a significant role of Madden-Julian oscillation (MJO) and El Niño in the frequency of occurrence of TCs (see [13]). In certain scenarios, equatorial Rossby waves (ER), mixed Rossby-gravity waves (MRG), Kelvin waves and easterly waves also influence the tropical cyclogenesis [14]. However, equatorial Kelvin waves do not appear to play a major role in tropical cyclogenesis. Several hurricanes in the North Atlantic form from African waves, MJO has a significant role in tropical cyclogenesis in North Pacific and the formation of cyclonic storms in the northwest Pacific is associated with MRG waves [13, 15]. These waves enhance the local conditions for the genesis of TCs by increasing upward motion, convection and the low-level vorticity by altering the local vertical shear pattern. The larger-scale waves, such as the MJO and ER, can also alter the mean zonal wind in large spatial and temporal scales in order to influence the mean flow.

The active phase of MJO is generally found over Indian Ocean, the maritime continent, and western pacific [13], which seem to play a major role in regulating the frequency of occurrence (usually increases) and formation of TCs in these regions. MJO increases the westerly wind which blows from west to east and its active phase through the region increases convective activity. During El Niño events, the atmospheric response to SST anomalies (SSTA) in the equatorial Pacific perturbs the Walker circulation [16]. The most common form of genesis occurs when they interact with the Asian monsoon. However, such type of interaction is still not studied well though few studies emphasized the role of El Niño or El Niño Southern Oscillation (ENSO) in TC formation over the Bay of Bengal (BOB) region indicating a decrease in the number of TCs [17].

2.2. Structural evolution

The general structure of the TC can be understood through the visualization of vertical cross-section of a mature TC as depicted in **Figure 1**, which consists of eye, eyewall and rainbands. The centre of the structure signifies the low-pressure cyclone eye, where a strong downward flow occurs indifferent to the immediate neighbouring updrafts. However, subsidence is also visible alongside the updrafts in the neighbourhood eyewall region away from the eye. The appearance of eye, its growth, intensification of eyewall and disappearance of eye are described in this section.

The life cycle of the TC is shown in **Figure 2(a)**, where the inner and outer cores of a TC are considered besides its intensification in order to depict the strengthening and weakening. **Figure 2(b)** depicts the different stages of TC life period including genesis, development, mature stage and dissipation by considering the evolution of TC Phailin (2013) in BOB region as an example to the illustration shown in **Figure 2(a)**.

In the intensification period (or phase 1), the momentum from outer core towards the inner one helps in strengthening the 700 hPa wind field and subsequently it helps in the eye wall cloud formation [18]. Prior to the appearance of eye, the intensification process is quite slow (at a rate ~8 hPa/day). The increase in maximum wind field is ~5 $ms^{-1}day^{-1}$ and the outer core strengthens at a rate of ~2 $ms^{-1}day^{-1}$. Gradually, when the eye appears, the central pressure is about 987 hPa and the rate of intensification increases by ~250 times, at a rate of about 20 hPa/day. The rapidly deepening cyclone (at a rate ~42 hPa/day) supports an earlier eye formation. During the filling phase, the central pressure starts rising by drawing momentum through the outer core and strengthening the outer core's wing.

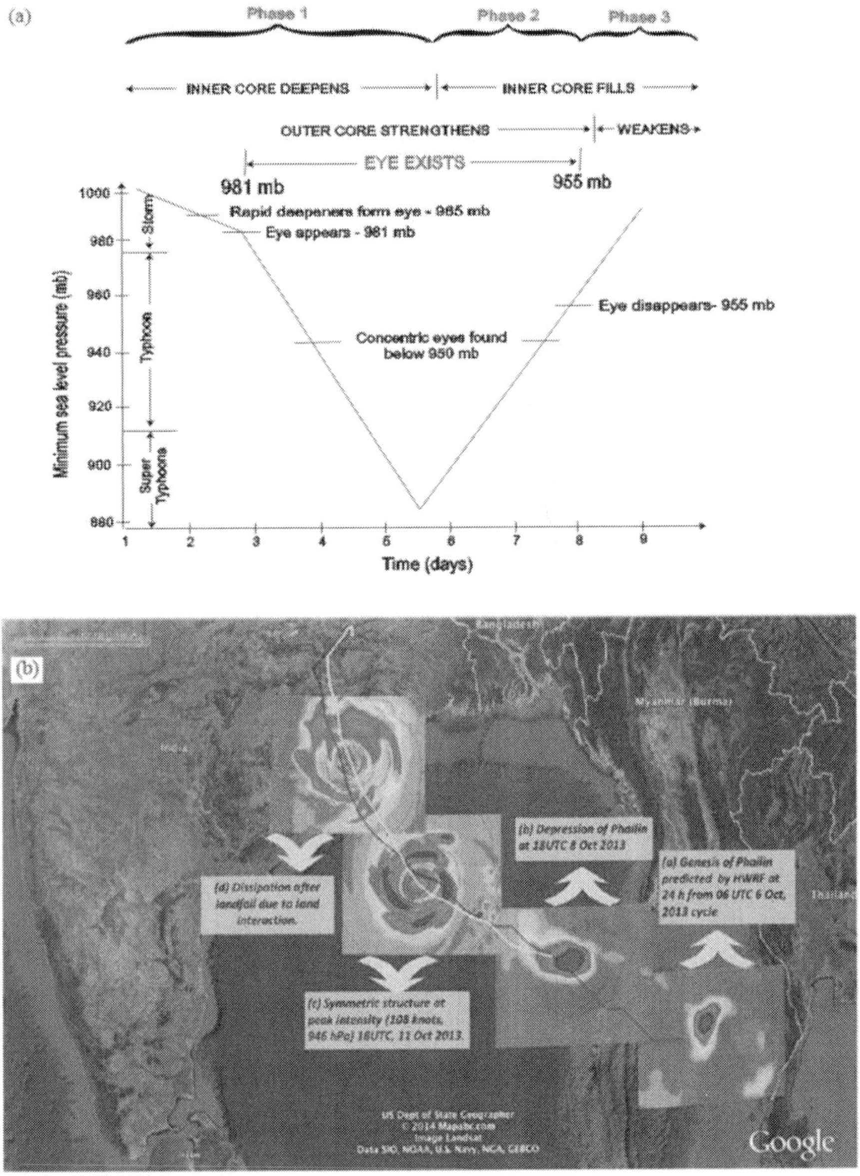

Figure 2. (a) Conceptual rendering from the main events in the life cycle of a tropical cyclone [18] and (b) different stages of tropical cyclone Phailin formed over Bay of Bengal [80].

The phase 2 is usually marked by the strengthening of outer core wind (similar to the stage (c) of TC Phailin shown in **Figure 2(b)**), whereas the inner core wind diminishes. During this phase, the eye expands and the filling of the inner core continues through the inflowing air towards the cyclone centre. During the filling phase, the inertial stability of the outer core is twice as large as that of the deepening stage making the outer core rigid for the inflowing air. Gradually the expansion of the eye ceases and the central core fills. It is important to note that the longer a cyclone spends in phase 2, stronger the outer radius will be and the radius of the damaging winds expand as long as the eye exists.

During phase 3, the outer wind starts weakening (e.g. stage (d) of TC Phailin shown in **Figure 2(b)**) with the disappearance of eye. Once the eye is vanished, the inflow of angular momentum ceases that was responsible for the strengthening of the outer core and from where the decay of outer radius low level wind field begins. These characteristics are valid for the cyclones which do not suffer landfall because the landfall would erode the wind field irrespective of the appearance of eye.

Though the basic principles of structural evolution may hold good for the TCs occurring in NIO basin (refer to **Figure 2(b)** for different stages of the TC Phailin), the formation of a distinguished eye structure may always not be feasible. A distinguished eye may be seen in case of a very severe cyclonic storm in this basin during phase 2. However, an explicit analysis in this direction is not available in literature for the NIO basin even though few studies like [19] computed the radius of maximum wind seen in case of TC intensification.

2.3. Propagation

TCs generally originate in tropics and thereafter, travel westward [20, 21] or turn poleward and recurve towards eastward direction [21, 22] or suffers extratropical transition over land or water [23] before dissipation. If a time scale of 1–3 weeks is considered, then the evolution of Rossby wave train significantly influences the track of a TC. Across the subtropical regions, under the influence of synoptic scale ridging, the TCs tend to move more westerly, and under the influence of synoptic scale trough, TCs tend to recurve into the mid-latitude [24]. On a seasonal scale, it is seen that over the Indian Ocean, the advancement of monsoon has a considerable impact on TCs' growth and their track [25].

In principle, TCs move under the influence of its surrounding environment. When the easterlies are added with the wind at certain level from the storm, the resulting effect forces the system to move in a westward direction [26]. Since the winds are not constant with height, it complicates the movement. The 'β effect' or 'β drift' pushes the cyclone towards the northwest direction in the northern hemisphere. It superimposes a weak northwest ward (southwest ward) steering current upon the TC in the northern (southern) hemisphere.

Apart from the factors mentioned earlier, the wind shear around anti-cyclonic flow at the top of the TCs also impacts their movement and can influence the track as much as the 'β drift'. There is a more complex phenomenon which influences the motion of a cyclone, known as 'Fujiwhara effect' [27]. Fujiwhara interaction describes the mutual rotation of two vortices about a common centre [28]. This centre typically refers to the mass weighted centroid of the two vortices, if they are of equal strength. In the presence of the β effect, the two vortices rotate around each other relative to the centre of rotation. This centre of rotation is not fixed and, instead, moves northwest ward in response to the 'β effect'. 'Fujiwhara effect' is noticed over other basins of the world including Atlantic, but is not applicable for TCs formed over NIO.

2.4. Dissipation

The most common way of dissipation of a TC is its landfall. When the storm moves over land, it deprives itself from warm water and the available moisture over ocean. Consequently, it is deprived from the energy source and the warm core with thunderstorms near the centre turns into a remnant low-pressure area due to quick loss of energy. Weakening can also occur if it encounters a vertical wind shear that causes the heat engine and convection shift away from the centre. The rate of power dissipation of TCs can be computed [29] as

$$E_D = C_D \rho v^3$$

where E_D is the rate of energy dissipation per unit time per unit horizontal surface area, v defines the wind speed, ρ is for air mass density, and 'C_D' is the drag coefficient that depends upon the surface irregularities. Since the power dissipation in TCs is proportional to the cube of its wind velocity, the severity can be computed as the cumulative sum of the cube of the wind velocity over time according to the above equation.

3. Role of ocean in genesis and intensification

There are two sources which are capable of changing the TC intensity, one is internal variability and other one is environmental interaction. One important aspect of later source is the interaction between the ocean and the storm system. Usually TC is regarded as the most forceful case in air-sea interaction studies where energy from the warm ocean waters is delivered via surface heat flux [30]. The ocean response is quite sensitive to the surface drag coefficient. Emanuel [31] used a simple numerical model to establish the progress of hurricane intensity. Their findings advocate that in most cases, the intensity depends on three factors, viz. initial intensity of cyclone, thermodynamic state of atmosphere through which the cyclone propagates and finally the heat exchange with the upper layer of the ocean underlying the core of the cyclone. Rapid intensification of TC is noticed when it passes over the deep upper ocean mixed layer and that upper ocean thermal structure plays a significant role in the intensification process [32–34]. Sutyrin [35] performed simulations with a coupled model of the oceanic and atmospheric boundary layers and concluded that the interaction is strong enough to change the supply of heat and moisture fluxes from the ocean into the atmosphere significantly within few hours of the formation of the storm and consequently, influence the TC intensity.

The intensity of TC increases with increase in SST and upper ocean heat content [36]. The positive feedback occurs when genesis and intensification happens. During this phase, the evaporation from the ocean surface stimulates surface wind that subsequently increases the moisture supply and consequently increasing the latent heat that is further utilized to drive the circulation. As a negative feedback, the decrease in SST results in the decrease in total heat flux (sum of latent heat and sensible heat), resulting in decrease in intensity of the storm. Besides these interactions, some of the mechanical energy supplied by the TC is dispersed laterally and vertically by the internal inertia-gravity waves with time [37].

On the other hand, the intensification of TC depends not only on SST but also on subsurface ocean thermal structure also considered as an important predictor for the TC intensification (e.g. see [38–40]). In the changing climate scenario, SST plays a bigger role during pre-monsoon season as compared to the post-monsoon period for governing TC activity over NIO region [41]. In contrast, the same may not be valid for other basins including North Atlantic Ocean, where an increasing trend in correlation between SST and TC power dissipative index is observed [42]. The influence of the changing climate on the TC genesis and intensification in the NIO region may therefore not be limited to the analysis relating SST only.

4. Numerical modelling of tropical cyclones

A significant number of studies regarding TC propagation, track prediction, time and place of landfall and intensity of the storm are carried out for several ocean basins including NIO. Considerable improvements in predicting the TCs are also achieved till date. In view of these, this section highlights the recent developments regarding TC predictability over NIO region and the current scenario.

4.1. Model predictability

Various regional models such as GFDL (USA), ALADIN (France), Quasi-Lagrangian Limited Area Model or QLM (India), MM5 (USA), etc. are used for TC research and operational forecasting purpose. Apart from these, the Eulerian-mass-based dynamical core of Weather Research and Forecasting (WRF) model, designed as the successor to MM5 is also used to predict TCs. The variants of WRF regional model are Advanced Research WRF or ARW and non-hydrostatic mesoscale model or WRF-NMM. Though these numerical models are quite capable for real-time predictions in regional scale, they need appropriate initial and boundary conditions from global models. For example, a recent study carried out by Kumar [43] discusses about the impact of European Centre for Medium-Range Weather Forecasts (ECMWF), National Centers for Environmental Prediction (NCEP) and National Centre for Medium Range Weather Forecasting (NCMRWF) global model analysis on the WRF model forecast for TC prediction over Indian region. This study indicates some of the inherent limitations of such global analyses data sets including the consideration of few fundamental aspects like that of the middle tropospheric humidity profiles those are important for TC genesis. Another limitation of such data sets is their horizontal resolution though recent advancements have made availability of some of the usable global analyses for the desired purpose with higher spatial resolutions up to 0.25°.

Since NWP models are equipped with real-time prediction capability, they are being used increasingly for the TC prediction over NIO region as well. Some of the numerical models and their skills are discussed here. For instance, QLM regional model was adopted by Prasad [44] for cyclone track prediction over NIO region and found the performance to be reasonable. The recurvature of the cyclones were also well predicted. However, the model performance for TC intensity prediction was not satisfactory. Another notable study by Mohanty et al. [45] used MM5 to simulate Orissa (Odisha) super cyclone (1999) for predicting track, intensity, mean sea level pressure and associated precipitation. Though such types of studies were able to improve the prediction of several relevant parameters including TC tracks, they were not so successful in predicting the intensity accurately like the studies performed using QLM. Similarly, some recent studies used three variants of the next-generation mesoscale WRF model (i.e. ARW, WRF-NMM, and Hurricane Weather Research and Forecasting Model or HWRF) for TC research and operational purpose as well [51, 53, 56, 57, 66]. It may be noted that ARW uses Arakawa C-grid staggering while WRF-NMM and HWRF use Arakawa E-grid. All of the WRF model variants use terrain following co-ordinate system and specific physical parameterizations. Since several modelling features in WRF are quite advanced (e.g. moving nest feature in HWRF) as compared to MM5, it is expected that at least one or more variants of it would show better performance for TC prediction over NIO region. Extensive research in this direction using ARW suggests some significant improvements in predicting the tropical cyclogenesis and cyclone tracks [10, 46–54]. However, it is noticed that improvement in prediction of TC intensity is found to be slower than that of track [51, 55].

A comparison study among MM5, WRF-ARW and WRF-NMM for very severe cyclone Mala (2006) developed over BOB found that ARW could simulate the TC intensity in terms of minimum central pressure and maximum sustainable wind with better accuracy [56]. However, MM5 simulated a more rapidly intensified storm and delayed landfall and WRF-NMM failed to simulate the intensity of the storm properly. On the other hand, WRF-NMM predicted TC track more accurately as compared to ARW and MM5. The TC Mala when simulated using HWRF with different initial conditions, the track error was found to be ~200 km and the intensity prediction was reasonably good for some considered initial conditions though the amount and spatial distribution of rainfall was well simulated by the model [57]. In order to improve the predictability, appropriate nesting technique, horizontal and vertical resolutions as well as physical parameterizations are considered [59, 68] besides data assimilation [60]. In view of these aspects, the HWRF system is now implemented at IMD along with the already operational ARW model for forecasting of TCs over NIO basin. As part of the Forecast Demonstration Project (FDP) conducted by IMD, it is analysed that the performance of ARW without data assimilation is reasonable over BOB [61]. Its performance improves when available observations are assimilated. Similar is the case with WRF-NMM. On the other hand, HWRF is capable of simulating rapid intensification of TCs over NIO region due to its improved vortex relocation and initialization procedures [49].

The high-resolution mesoscale modelling systems provide better guidance for TC forecast up to 72 h over NIO region [61]. They require high-resolution global analyses data sets for appropriate initial and boundary conditions in order to bring in large-scale boundary forcing

[62]. In order to reduce model errors, the initial and boundary conditions can be improved by adopting appropriate data assimilation techniques by incorporating the conventional, radar and satellite observations before running the model [61]. Thus, these aspects need special attention as far as predictability of TCs over NIO region is concerned.

4.2. Role of physical parameterizations

The physical parameterizations which include cumulus convection, surface fluxes of heat, moisture, momentum and vertical mixing in the planetary boundary layer play an important role in determining structural development, intensification and movement of TCs [10, 46, 48, 50, 53, 58, 63–65]. A number of studies emphasized upon these aspects during the past three decades. For the simulation purpose, they use the previously mentioned models (see Section 5.1). Most of these studies conduct simulations over a particular ocean basin. For instance, Osuri et al. [50] conducted a systematic study on customization of ARW model considering several physical parameterization schemes for the simulation of five TCs over NIO region. The study found that the combination of Yonsei University (YSU) planetary boundary layer (PBL) parameterization with KF convection scheme provided a better prediction for structural characteristics, intensity, track and rainfall. Similar results were also achieved by several studies including that of [10, 46, 48]. Thus, most of the studies (including [65]) found the performance of KF scheme to be better for the prediction of TCs over NIO region. However, recent studies by Kanase and Salvekar [53] obtained that the Betts-Miller-Janjic (BMJ) convection scheme performs better as compared to other parameterizations in the group although the study also favoured using YSU PBL physics. On the other hand, it found that WRF single-moment (WSM)-6 microphysics better represents mid-tropospheric heating as compared to WSM-3 favouring better intensity simulation.

Though HWRF has not been extensively used for sensitivity studies with respect to physical parameterizations for simulation of TCs over NIO region, its primitive variant WRF-NMM was used in recent past by some of the researchers. For example, studies by Pattanayak et al. [66] found that the combination of Simplified Arakawa-Schubert (SAS) convection, YSU PBL, Ferrier microphysics and NMM land-surface parameterization schemes in WRF-NMM performs better in predicting track and intensity of TC Nargis (2008) over BOB. Therefore, an extensive evaluation of HWRF is needed in order to determine the combination of physical parameterizations that performs better for TC prediction over NIO region before it is adopted for the operational forecasting purpose.

4.3. Significance of grid resolution

The grid resolution of a model also impacts the TC prediction [51, 58, 59, 67]. However, there are very few studies available relating to the impact of grid resolution on TC prediction over NIO region. One of the notable studies by Rao [68] evaluated the impact of horizontal resolution and the advantages of the nested domain approach in the prediction of Orissa (Odisha) super cyclone intensification and movement by using MM5 model. Results from this study indicate that the enhancement of resolution produces higher intensity but does not influence the track of the storm. The nested experiments produced cyclone track closely agreeing with the observations, while the single domain based simulations show the deviation of the track towards north. A more recent study by Osuri et al. [51] found that the use of high resolutions in operational ARW model improves the prediction of recurving TC tracks and their intensity. In a climatological framework, Community Atmospheric Model or CAM showed sensitiveness to the prediction of more number of intensified tropical cyclones over most of the global basins including NIO. Further, it also found that the duration of tropical storms would be much larger in high resolutions simulations. Thus, it is realized that the model horizontal grid resolution impacts significantly the TC track, intensity and duration besides other relevant meteorological parameters.

4.4. Significance of data assimilation

Most of the times, the use of data assimilation techniques in TC simulations helps in improving the model predictability. For this purpose, satellite-based observations, aircraft measurements and radar data are used besides the conventional data sets. The widely used data assimilation techniques are primarily based on either ensemble Kalman filter (EnKF) or variational techniques (3DVAR or 4DVAR). Most of the studies related to TC simulation were done using variational data assimilation techniques for improving the TC prediction over NIO region. For example, the studies such as [52, 69–71] used 3DVAR techniques for assimilating satellite, radar and conventional measurements for improving the initial and boundary conditions of MM5 and ARW mesoscale models in order to better predict TC structure, track, intensity and associated relevant meteorological variables including rainfall. In some situations, the improvement was not significantly noticed. For instance, the studies by Singh et al. [70] found that assimilation of SSM/I wind speed data resulted in simulating weak intensity and failed to make an impact on track prediction.

Although there are no significant studies related to the use of 4DVAR and EnKF techniques for simulating NIO TCs, there are literatures, which demonstrate the usage of four dimensional data assimilation (FDDA) nudging technique in order to improve the ARW model predictability. For example, [71–73] used FDDA nudging technique in order to improve ARW initial and boundary conditions for the simulation of several TCs over NIO region those occurred during 2007–2010. These studies primarily emphasized upon TC track and intensity forecasts. While some of them reported remarkable improvements in track prediction and landfall position with either 12- or 18-h of nudging yielding maximum impact [72, 74], some others noticed relatively less impact of FDDA observational nudging on intensity prediction [73].

5. Tropical cyclone climatology over NIO region

Since hundreds of years, the Indian Ocean is a breeding basin for disastrous TCs associated with heavy rainfall, torrential wind and storm surges. The cyclones in 1970 and 1991 caused a loss of more than 400,000 lives. During the Odisha super cyclone (1999), more than 10,000 lives were lost and a destruction of 1.9 million houses occurred in 14 districts. Recently, Nargis (2008) caused ~1 40,000 deaths in Myanmar. In 2015, cyclonic storm Komen caused a heavy loss throughout Bangladesh, Myanmar, northeast India and eastern parts of India although the loss of lives was very few as compared to previous cases because of the improvement in TC predictability. This was also realized in case of Phailin (2013) and Hudhud (2014).

TCs usually form over NIO basin in two seasons, that is, pre-monsoon (March-April-May) and post-monsoon (October–November–December) period. In total, about 1108 numbers of cyclonic systems are formed over NIO region (includes both BOB and Arabian Sea, AS) during 1891–2015. It includes depressions (or D), cyclonic storms (or CS) and severe cyclonic storms (or SCS). However, the cyclonic systems do not form each month of every year. If the average monthly distribution of these three types of cyclonic systems (**Figure 3**) is analysed, it is evident that maximum number of cyclones occur between the months of May to December. Maximum numbers of depressions are formed in August. Maximum numbers of CS are formed in the month of October, while November is the most favourable month for the formation of SCS. Though the number of total cyclonic systems in May is relatively less, ~48.7% of cyclonic disturbances are transformed to very severe cyclonic storms. However, this transformation is found to be 43.9 and 41.7%, respectively, in the months of April and November. Annually the probability of intensification of depression to CS is ~44.8%, depression to SCS is ~21.3% and the probability of intensification of CS to SCS is ~47.5%.

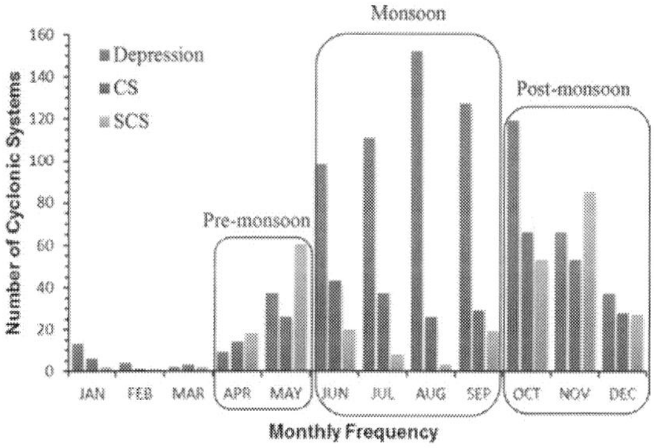

Figure 3. Monthly frequency of cyclonic disturbances in North Indian Ocean region during 1891–2015. Here depression signifies the low-pressure systems which do not transform to cyclonic storms; CS is for the cyclonic storms and SCS represents the severe cyclonic storms.

BOB contributes about 75% of TCs during cyclone seasons (pre- and post-monsoon periods) and the AS contributes ~25% [75]. The possible reason could be that BOB is generally more stratified than AS because its upper-ocean part is relatively warmer resulting in higher SST. In addition, low flat coastal terrain and funnel shape, shallow water of BOB [76], monsoonal wind (trough), more middle tropospheric moisture availability and lower tropospheric westward travelling disturbances such as easterly waves (often serve as the 'seedling' circulations) play roles in generating more number of cyclonic systems over BOB. Most of the monsoon troughs generated because of re-intensification of westerly propagating disturbances or from in situ depressions help in the formation of cyclonic systems over this region as well. Boreal summer intraseasonal oscillation (BSISO) also modulates the topical cyclogenesis over BOB [77], and it may be noted that the genesis potential index is high during the active phase of the BSISO.

The studies like that of [4] indicate that under the global warming scenario, the number and proportion of cyclones reaching SCS are increasing in almost all basins of the world especially indicating the impact of climate change. **Figure 4** shows the decadal variation of cyclonic disturbances and CSs over NIO, that is, over BOB and AS. It is clear from the curve that there is a significant decreasing trend in the number of cyclonic disturbances and CS. When the number of SCS are analysed, it shows a slight increase or may be considered as a constant trend in decadal scale (**Figure 4**). During 1961–1970 and 1971–1980, there was most number of SCS. Besides El-Nino Southern Oscillation (ENSO), MJO (Madden-Julian Oscillation) and IOD (Indian Ocean Dipole) may also play appreciable role in modulating the TC activity over NIO region [13, 16, 17, 77].

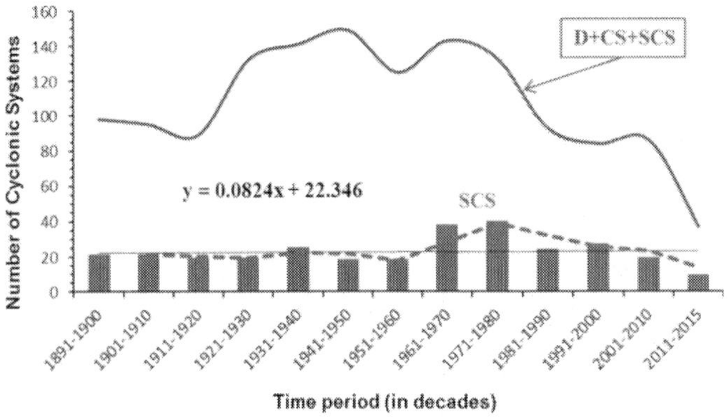

Figure 4. Variation of decadal frequency of cyclonic disturbances or depressions (D), cyclonic storms (CS) and severe cyclonic storms (SCS) over NIO region (smooth curved line). The bar diagrams represent SCS during 1891–2015. The dotted line indicates the moving trend and line shows the linear trend.

For the past three decades, the number of SCS has somehow decreased to a considerable value (**Figure 4**). However, Mohanty et al. [75] demonstrated that there is a considerable increase of SCS by about 65% during the warming period 1951–2007 by analysing the genesis and intensity of TCs over NIO basin in yearly scale. In the southern sector of BOB, a considerable increase of ~71% in SCS is found in post-monsoon season. Rate of dissipation of SCS over BOB is also significantly reduced besides increase in mean SST in the warming scenario and these features contribute to increase in the number of SCS over NIO. In the western sector of AS, a significant increase in SCS is also observed in the warming conditions. Therefore, the intensity of the SCS is increasingly becoming significant in the changing climate scenario. When the 'T Numbers' of the cyclones are analysed in satellite era, it is found that the Odisha super cyclone (1999) was the strongest recorded CS in the NIO basin during 1990–2015.

http://dx.doi.org/10.5772/64333

Analysing the track of cyclones over BOB and AS from e-atlas available at IMD, New Delhi, it is observed that most of the cyclonic systems developing over the NIO basin move in a northwesterly direction. However, there are cases of recurvature towards the northeast or east to southwest. The frequency of recurvature is higher towards the northeast compared to southwest or east. The probability of recurvature is higher over the AS when the system moves to the north of 15°N increasing the possibility of landfall over Gujarat coast. Over BOB, there is no such preferred latitude/longitude for the recurvature prospects. On the other hand, the probability of recurvature towards northeast region is higher during the pre-monsoon season.

Out of 1108 cyclones formed during last 124 years, 751 (68%) have crossed east coast of India, 214 (19.31%) Bangladesh, 57 (5.18%) Myanmar, 63 (5.68%) west coast of India and 26 (2.3%) numbers of cyclones crossed the coastal regions between India and Pakistan affecting the economy of both the countries. According to studies by Tyagi et al. [78], over 60% of TCs formed over BOB suffer landfall in different parts of east coast of India, 30% strike coasts of Bangladesh and Myanmar and about 10% dissipate over the sea itself. The differences in observed percentages are because of the obvious reason, that is, consideration of different time periods. However, it is evident that NIO basin is quite significant in view of the TC occurrence and highly populated and economically growing south Asian region.

6. Ongoing activities and possible recommendations for future

In order to improve the prediction of TC predictability over BOB region, the modernization of the observational system is being carried out by IMD, which includes setting up of two clusters of surface meso-meteorological networks: one along the coasts of Odisha-West Bengal and the other around Andhra Pradesh coasts [2]. About 443 numbers of existing automatic weather station (AWS) are there set up in different states of India. For NIO basin, it is considered very important to acquire weather reconnaissance aircraft facility to provide information on environmental winds and thermodynamical structures in the inner core region of TCs. The FDP (2008) is an attempt in this direction to determine the possible improvements in track and landfall predictions by using aircraft data.

The programmes named as STORM and PRWONAM are carried out with the support of Ministry of Earth Sciences (MOES) and Department of Science and Technology. MOES is also involved in strengthening of the deep ocean and met-ocean buoys network. In addition, IMD has established high wind speed recorder systems, S-band Doppler radars and Global Positioning System (GPS) equipment along the coastal areas of India [79]. Under the Indo-French collaboration, Oceansat-II (was functional till 2014) and MEGHA-TROPIQUES satellite with capability of repeated scanning over BOB region are/were functional to provide data related to sea surface winds, clouds, humidity, temperature, rainfall and radiation. The earth receiving stations for METOP and MODIS satellite data have been installed at IMD. Products like cloud motion vector (CMV), water vapour wind (WVW), out-going longwave radiation (OLR), quantitative precipitation estimate (QPE), Sea Surface Temperature (SST), upper tropospheric humidity (UTH) and cloud top temperature (CTT) are derived from other satellites including KALPANA-1 and INSAT-3D.

Several research institutes such as National Centre for Medium Range Weather Forecasting, Noida; Indian National Centre for Ocean Information Services, Hyderabad; Indian Space Research Organization (ISRO), Air Force and academic institutes including IITs (Indian Institute of Technology), NITs (National Institute of Technology), universities contribute towards providing their valuable input through academic research regarding various aspects of TC activity over NIO region. With these inputs and in-house research and development, IMD has been able to strengthen its capability in recent past, both from numerical modelling as well as observational point of view by taking into account both in situ and satellite measurements.

Despite increased capability for TC prediction over NIO region, few aspects still need to be addressed. Those key areas include accuracy in track prediction, time and place of landfall, accurate storm surge prediction and improving the intensity predictability. In addition, the changes in tropical cyclogenesis need to be understood in the changing climate scenario. It is because the severity of TCs is found to be increasing in the warming environment [75]. The improvement in numerical model predictions can be done by improving physical parameterization schemes, incorporating observations from different sources including those from satellites and radars in the model initial and boundary conditions through appropriate data assimilation techniques and considering improved SSTs. In addition, better disaster management need to be done alongside in order to reduce the loss of lives and properties.

Author details

Kasturi Singh, Jagabandhu Panda*, Krishna K. Osuri and Naresh Krishna Vissa

*Address all correspondence to: pandaj@nitrkl.ac.in

Department of Earth and Atmospheric Sciences, National Institute of Technology Rourkela, Odisha, India

References

[1] Longshore, D. (2008). Encyclopedia of hurricanes, typhoons and cyclones, new edition, chapter A to Z Entries, p. 468. Facts on File.

[2] Mohapatra, M., Singh, R., Ray, K., Kotal, S. D., Goel, S., Singh, C., Kumar, N., Ashrit, R. G., Balachandran, S., Rathore, L. S., Bandyopadhyay, B. K., Mohanty, U. C., Osuri Krishna K., Sikka, D. R., Basu, S., Thampi, S. B., Ramanan, S. R., & Rao, K. R. (2015). Forecast Demonstration Project (FDP) for improving track, intensity and landfall of Bay of Bengal tropical cyclones, implementation of pilot phase, 2014: a report. Report No.: FDP/TCR/1/2015.

[3] Bahinipati, C. S. (2014). Assessment of vulnerability to cyclones and floods in Odisha, India: a district-level analysis. Current Science, 107, 1997–2007.

[4] Webster, P. J., Holland, G. J., Curry, J. A., & Chang, H. R. (2005). Changes in tropical cyclone number, duration, and intensity in a warming environment. Science, 309, 1844–1846.

[5] Palmen, E. (1956). A review of knowledge on the formation and development of tropical cyclones. Proceedings of the Tropical Cyclone Symposium, Bureau of Meteorology, Brisbane, Australia, 213–232.

[6] Frank, W. M. (1987). Tropical cyclone formation (Chapter 3), a global view of tropical cyclones, WMO Bangkok, Thailand textbook, Printed by Dept. of Chicago, 53–90.

[7] Zehr, R. (1992). Tropical cyclogenesis in the western North Pacific. NOAA Technical Report NESDIS 16, 181 pp. (available from NESD1S, Washington, DC or CIRA, Colo. State Univ., Ft. Collins, CO).

[8] Gray, W. M. (1975). Tropical cyclone genesis. Dept. of Atmos. Sci. Paper No. 234, Colo. State Univ., Ft. Collins, CO, 121 pp.

[9] Gray, W. M. (1979). Hurricanes: their formation, structure and likely role in the tropical circulation. In: Supplement of Meteorology over the Tropical Oceans. Published by RMS, James Glaisher House, Grenville Place, Bracknell, Berkshire, RG 12 1BX, D. B. Shaw (ed.), 155–218.

[10] Panda, J., Singh, H., Wang, P. K., Giri, R. K., & Routray, A. (2015). A qualitative study of some meteorological features during tropical cyclone PHET using satellite observations and WRF modeling system. Journal of the Indian Society of Remote Sensing, 43, 45–56.

[11] Anthes, R. A. (1974). The dynamics and energetics of mature tropical cyclones. Reviews of Geophysics, 12, 495–522.

[12] Marks Jr, F. D. (2003). State of the science: radar view of tropical cyclones. In Radar and Atmospheric Science: A Collection of Essays in Honor of David Atlas, American Meteorological Society, pp. 33–74, DOI:10.1175/BAMS-87-11-1523.

[13] Tsuboi, A., & Takemi, T. (2014). The interannual relationship between MJO activity and tropical cyclone genesis in the Indian Ocean. Geoscience Letters, 1, 1–6.

[14] Frank, W. M., & Roundy, P. E. (2006). The role of tropical waves in tropical cyclogenesis. Monthly Weather Review, 134, 2397–2417.

[15] Frank, W. M., & Clark, G. B. (1980). Atlantic tropical systems of 1979. Monthly Weather Review, 108, 966–972.

[16] Sumesh, K. G., & Kumar, M. R. (2013). Tropical cyclones over north Indian Ocean during El-Niño Modoki years. Natural Hazards, 68, 1057–1074.

[17] Girishkumar, M. S., & Ravichandran, M. (2012). The influences of ENSO on tropical cyclone activity in the Bay of Bengal during October–December. Journal of Geophysical Research: Oceans, 117, DOI: 10.1029/2011JC007417.

[18] Weatherford, C. L. (1989). The structural evolution of typhoons. Atmospheric Science Paper, 0067-0340; no. 446.

[19] Mehra, P., Soumya, M., Vethamony, P., Vijaykumar, K., Nair, B., Agarvadekar, Y., & Harmalkar, B. (2015). Coastal sea level response to the tropical cyclonic forcing in the northern Indian Ocean. Ocean Science, 11, 159–173.

[20] Simpson, R. H. (1946). On the movement of tropical cyclones. Eos, Transactions American Geophysical Union, 27, 641–655.

[21] Elsberry R. L. (1995). Tropical cyclone motion. In: Global Perspectives on Tropical Cyclones, R. L. Elsberry (ed.). World Meteorological Organization, Geneva, Switzerland, Report No. TCP-38.

[22] Sampson, C. R., Jeffries, R. A., & Neumann, C. J. (1995). Tropical Cyclone Forecasters Reference Guide 4. Tropical Cyclone Motion (No. NRL/PU/7541-95-0010). Naval Research Lab, Monterey.

[23] Jones, S. C., Harr, P. A., Abraham, J., Bosart, L. F., Bowyer, P. J., Evans, J. L., & Sinclair, M. R. (2003). The extratropical transition of tropical cyclones: forecast challenges, current understanding, and future directions. Weather and Forecasting, 18, 1052–1092.

[24] Chan, J. C., & Gray, W. M. (1982). Tropical cyclone movement and surrounding flow relationships. Monthly Weather Review, 110, 1354–1374.

[25] Li, Z., Yu, W., Li, T., Murty, V. S. N., & Tangang, F. (2013). Bimodal character of cyclone climatology in the Bay of Bengal modulated by monsoon seasonal cycle. Journal of Climate, 26, 1033–1046.

[26] Emanuel, K. (2003). Tropical cyclones. Annual Review of Earth and Planetary Sciences, 31, 75.

[27] Fujiwhara, S. (1921). The natural tendency towards symmetry of motion and its application as a principle in meteorology. Quarterly Journal of the Royal Meteorological Society, 47, 287–292.

[28] Wang, Y., & Holland, G. J. (1995). On the interaction of tropical-cyclone-scale vortices. IV: Baroclinic vortices. Quarterly Journal of the Royal Meteorological Society, 121, 95–126.

[29] Emanuel, K.A., 1998: The power of a hurricane: an example of reckless driving on the information superhighway. Weather, 54, 107–108.

[30] Emanuel, K. A. (1986). An air-sea interaction theory for tropical cyclones. Part I: Steady-state maintenance. Journal of the Atmospheric Sciences, 43, 585–605.

[31] Emanuel, K. A. (1999). Thermodynamic control of hurricane intensity. Nature, 401, 665–669.

[32] Subrahmanyam, B., Murty, V. S. N., Sharp, R. J., & O'Brien, J. J. (2005). Air-sea coupling during the tropical cyclones in the Indian Ocean: A case study using satellite observations. Pure and Applied Geophysics, 162, 1643–1672.

[33] Vissa, N. K., Satyanarayana, A. N. V., & Kumar, B. P. (2012). Response of Upper Ocean during passage of MALA cyclone utilizing ARGO data. International Journal of Applied Earth Observation and Geoinformation, 14, 149–159.

[34] Vissa, N. K., Satyanarayana, A. N. V., & Prasad Kumar, B. (2013). Response of oceanic cyclogenesis metrics for NARGIS cyclone: a case study. Atmospheric Science Letters, 14, 7–13.

[35] Sutyrin, G. G., Khain, A. P., & Agrenich, E. A. (1979). Interaction of the boundary layers of the ocean and the atmosphere on the intensity of a moving tropical cyclone. Meteorology Girology, 2, 45–56.

[36] Balaguru, K., Taraphdar, S., Leung, L. R., & Foltz, G. R. (2014). Increase in the intensity of postmonsoon Bay of Bengal tropical cyclones. Geophysical Research Letters, 41, 3594–3601.

[37] Ginis, I., 1995: Interaction of tropical cyclones with the ocean. *in Global Perspective of Tropical Cyclones*, Chapter 5, Ed. R. L. Elsberry, Tech. Document WMO/TD 693, World Meteorological Organization, Geneva, Switzerland, 198–260.

[38] Goni, G. J., & Trinanes, J. A. (2003). Ocean thermal structure monitoring could aid in the intensity forecast of tropical cyclones. Eos Transactions American Geophysical Union, 84, 573–578.

[39] Lin, I. I., Chen, C. H., Pun, I. F., Liu, W. T., … Wu, C. C. (2009). Warm ocean anomaly, air sea fluxes, and the rapid intensification of tropical cyclone Nargis (2008). Geophysical Research Letters, 36 (3), L03817, DOI: 10.1029/2008GL035815.

[40] Wada, A., Usui, N., & Sato, K. (2012). Relationship of maximum tropical cyclone intensity to sea surface temperature and tropical cyclone heat potential in the North Pacific Ocean. Journal of Geophysical Research: Atmospheres, 117(D11), D11118, DOI: 10.1029/2012JD017583.

[41] Sebastian, M., & Behera, M. R. (2015). Impact of SST on tropical cyclones in North Indian Ocean. Procedia Engineering, 116, 1072–1077.

[42] Emanuel, K. (2005). Increasing destructiveness of tropical cyclones over the past 30 years. Nature, 436, 686–688.

[43] Kumar, P., Kishtawal, C. M., & Pal, P. K. (2015). Impact of ECMWF, NCEP, and NCMRWF global model analysis on the WRF model forecast over Indian Region. Theoretical and Applied Climatology, 9pp, DOI: 10.1007/s00704-015-1629-1.

[44] Prasad, K., & Rao, Y. R. (2003). Cyclone track prediction by a quasi-Lagrangian limited area model. Meteorology and Atmospheric Physics, 83, 173–185.

[45] Mohanty, U. C., Mandal, M., & Raman, S. (2004). Simulation of Orissa super cyclone (1999) using PSU/NCAR mesoscale model. Natural Hazards, 31, 373–390.

[46] Panda, J., Giri, R. K., Patel, K. H., Sharma, A. K., & Sharma, R. K. (2011). Impact of satellite derived winds and cumulus physics during the occurrence of the tropical cyclone Phyan. Indian Journal of Science and Technology, 4, 859–875.

[47] Deshpande, M., Pattnaik, S., & Salvekar, P. S. (2010). Impact of physical parameterization schemes on numerical simulation of super cyclone Gonu. Natural Hazards, 55, 211–231.

[48] Raju, P.V.S., Potty J., Mohanty U.C. (2011) Sensitivity of physical parameterizations on the prediction of tropical cyclone Nargis over the BoB using WRF model. Meteorology and Atmospheric Physics, 113, 125–137.

[49] Rao, D. B., & Tallapragada, V. (2012). Tropical cyclone prediction over Bay of Bengal: a comparison of the performance of NCEP operational HWRF, NCAR ARW, and MM5 models. Natural Hazards, 63, 1393–1411.

[50] Osuri, K. K., Mohanty, U. C., Routray, A., Makarand, A. K., & Mohapatra, M. (2012). Sensitivity of physical parameterization schemes of WRF model for the simulation of Indian seas tropical cyclones. Natural Hazards, 63, 1337–1359.

[51] Osuri, K. K., Mohanty, U. C., Routray, A., Mohapatra, M., & Niyogi, D. (2013). Real-time track prediction of tropical cyclones over the North Indian Ocean using the ARW model. Journal of Applied Meteorology and Climatology, 52, 2476–2492.

[52] Osuri, K. K., Mohanty, U. C., Routray, A., & Niyogi, D. (2015). Improved prediction of Bay of Bengal Tropical cyclones through assimilation of Doppler weather radar observations. Monthly Weather Review, 143, 4533–4560.

[53] Kanase RD and P. S. Salvekar, (2015). Impact of physical parameterization schemes on track and intensity of severe cyclonic storms in Bay of Bengal. Meteorology and Atmospheric Physics, 127, 537–559.

[54] Nadimpalli, R., Osuri, K. K., Pattanayak, S., Mohanty, U. C., Nageswararao, M. M., & Prasad, S. K. (2016). Real-time prediction of movement, intensity and storm surge of very severe cyclonic storm Hudhud over Bay of Bengal using high-resolution dynamical model. Natural Hazards, 81, 1771–1795.

[55] DeMaria, M., Sampson, C. R., Knaff, J. A., & Musgrave, K. D. (2014). Is tropical cyclone intensity guidance improving? Bulletin of the American Meteorological Society, 95, 387–398.

[56] Pattanayak, S., Mohanty, U. C., Rizvi, S. R., Huang, X. Y., & Ratna, K. N. (2008). A comparative study on performance of MM5 and WRF (ARW & NMM) models

in simulation of tropical cyclone over Bay of Bengal. Current Science, 95, 923–936.

[57] Pattanayak, S., Mohanty, U. C., & Gopalakrishnan, S. G. (2012). Simulation of very severe cyclone Mala over Bay of Bengal with HWRF modeling system. Natural Hazards, 63, 1413–1437.

[58] Gopalakrishnan, S. G., Goldenberg, S., Quirino, T., Zhang, X., Marks Jr, F., Yeh, K. S., & Tallapragada, V. (2012). Toward improving high-resolution numerical hurricane forecasting: Influence of model horizontal grid resolution, initialization, and physics. Weather and Forecasting, 27, 647–666.

[59] Goldenberg, S. B., Gopalakrishnan, S. G., Tallapragada, V., Quirino, T., Marks Jr, F., Trahan, S., & Atlas, R. (2015). The 2012 triply nested, high-resolution operational version of the Hurricane Weather Research and Forecasting Model (HWRF): track and intensity forecast verifications. Weather and Forecasting, 30, 710–729.

[60] Bernardet, L., Tallapragada, V., Bao, S., Trahan, S., Kwon, Y., Liu, Q., & Carson, L. (2015). Community support and transition of research to operations for the Hurricane Weather Research and Forecasting Model. Bulletin of the American Meteorological Society, 96, 953–960.

[61] Mohanty, U. C., Osuri, K. K., & Pattanayak, S. (2013). A study on high resolution mesoscale modeling systems for simulation of tropical cyclones over the Bay of Bengal. Mausam, 64, 117–134.

[62] Kumar, A., Done, J., Dudhia, J., & Niyogi, D. (2011). Simulations of Cyclone Sidr in the Bay of Bengal with a high-resolution model: sensitivity to large-scale boundary forcing. Meteorology and Atmospheric Physics, 114, 123–137.

[63] Anthes, R. A. (1982). Tropical cyclones: their evolution, structure and effects. Boston: American Meteorological Society, 41, 1.

[64] Gopalakrishnan, S. G., Marks Jr, F., Zhang, J. A., Zhang, X., Bao, J. W., & Tallapragada, V. (2013). A study of the impacts of vertical diffusion on the structure and intensity of the tropical cyclones using the high-resolution HWRF system. Journal of Atmospheric Sciences, 70, 524–541.

[65] Reddy, M. V., Prasad, S. S., Krishna, U. M., & Ra, K. K. (2014). Effect of cumulus and microphysical parameterizations on the JAL cyclone prediction. Indian Journal of Radio and Space Physics, 43, 103–123.

[66] Pattanayak, S., Mohanty, U. C., & Osuri, K. K. (2012). Impact of parameterization of physical processes on simulation of track and intensity of tropical cyclone Nargis (2008) with WRF-NMM model. The Scientific World Journal, 2012, DOI:10.1100/2012/671437.

[67] Gopalakrishnan, S. G., Marks Jr, F., Zhang, X., Bao, J. W., Yeh, K. S., & Atlas, R. (2011). The experimental HWRF system: A study on the influence of horizontal resolution on

the structure and intensity changes in tropical cyclones using an idealized framework. Monthly Weather Review, 139, 1762–1784.

[68] Rao, A. D., Joshi, M., Jain, I., & Ravichandran, M. (2010). Response of subsurface waters in the eastern Arabian Sea to tropical cyclones. Estuarine, Coastal and Shelf Science, 89, 267–276.

[69] Panda, J., & Giri, R. K. (2012). A comprehensive study of surface and upper-air characteristics over two stations on the west coast of India during the occurrence of a cyclonic storm. Natural Hazards, 64, 1055–1078.

[70] Singh, R., Pal, P. K., Kishtawal, C. M., & Joshi, P. C. (2008). The impact of variational assimilation of SSM/I and QuikSCAT satellite observations on the numerical simulation of Indian Ocean tropical cyclones. Weather and Forecasting, 23, 460–476.

[71] Srivastava, K., & Bhardwaj, R. (2014). Analysis and very short range forecast of cyclone "AILA" with radar data assimilation with rapid intermittent cycle using ARPS 3DVAR and cloud analysis techniques. Meteorology and Atmospheric Physics, 124, 97–111.

[72] Kanase, R. D., & Salvekar, P. S. (2013). Role of Four-Dimensional Data Assimilation on Track and Intensity of Severe Cyclonic Storms. ISRN Meteorology, 2013. Available online at: http://dx.doi.org/10.1155/2013/972713.

[73] Srinivas, C. V., Yesubabu, V., Hariprasad, K. B. R. R., Ramakrishna, S. S. V., & Venkatraman, B. (2013). Real-time prediction of a severe cyclone 'Jal'over Bay of Bengal using a high-resolution mesoscale model WRF (ARW). Natural Hazards, 65, 331–357.

[74] Yesubabu, V., Srinivas, C. V., Ramakrishna, S. S. V. S., & Prasad, K. H. (2014). Impact of period and timescale of FDDA analysis nudging on the numerical simulation of tropical cyclones in the Bay of Bengal. Natural Hazards, 74, 2109–2128.

[75] Mohanty, U. C., Osuri, K. K., Pattanayak, S., & Sinha, P. (2012). An observational perspective on tropical cyclone activity over Indian seas in a warming environment. Natural Hazards, 63, 1319–1335.

[76] McBride, J. L., & Fraedrich, K. (1995). CISK: A theory for the response of tropical convective complexes to variations in sea surface temperature. Quarterly Journal of the Royal Meteorological Society, 121, 783–796.

[77] Kikuchi, K., & Wang, B. (2010). Formation of tropical cyclones in the northern Indian Ocean associated with two types of tropical intraseasonal oscillation modes. Journal of Meteorological Society of Japan, 88, 475–496, doi:10.2151/jmsj.2010-313.

[78] Tyagi, A., Bandyopadhyay, B. K., & Mohapatra, M. (2010). Monitoring and Prediction of Cyclonic Disturbances Over North Indian Ocean by Regional Specialised Meteorological Centre, New Delhi (India): Problems and Prospective. In Indian Ocean Tropical Cyclones and Climate Change, Y. Charabi (ed.), Springer Netherlands, pp. 93–103.

[79] Report on cyclonic disturbances over North Indian Ocean during 2012. (2013). Indian Meteorological Department, RSMC-Tropical Cyclones Report No/2013.

[80] Mohanty, U. C., Osuri, K. K., Tallapragada, V., Marks, F. D., Pattanayak, S., Mohapatra, M., Rathore, L. S., Gopalakrishnan, S. G., & Niyogi, D. (2015). A great escape from the Bay of Bengal " super sapphire–Phailin" tropical cyclone: a case of improved weather forecast and societal response for disaster mitigation. Earth Interactions, 19(17), 1–11.

Chapter 10

An Operational Statistical Scheme for Tropical Cyclone-Induced Rainfall Forecast

Qinglan Li , Hongping Lan , Johnny C.L. Chan ,
Chunyan Cao , Cheng Li and Xingbao Wang

Additional information is available at the end of the chapter

http://dx.doi.org/10.5772/64859

Abstract

Nonparametric methods are used in this study to analyze and predict short-term rainfall due to tropical cyclones (TCs) in a coastal meteorological station. All 427 TCs during 1953–2011, which made landfall along the Southeast China coast with a distance less than 700 km to a certain meteorological station, Shenzhen, are analyzed and grouped according to their landfalling direction, distance, and intensity. The corresponding daily rainfall records at Shenzhen Meteorological Station (SMS) during TCs landfalling period (a couple of days before and after TC landfall) are collected. The maximum daily rainfall (R24) and maximum 3-day accumulative rainfall (R72) records at SMS for each TC category are analyzed by a nonparametric statistical method, percentile estimation. The results are plotted by statistical boxplot, expressing in the probability of precipitation. The performance of the statistical boxplots was evaluated to forecast the short-term rainfall at SMS during the TC seasons in 2012 and 2013. The results show that the boxplot scheme can be used as a valuable reference to predict the short-term rainfall at SMS due to TCs landfalling along the Southeast China coast.

Keywords: tropical cyclone, rainfall forecast, nonparametric method, boxplot

1. Introduction

Tropical cyclones (TCs) are among the most destructive natural phenomena. TCs often bring about strong wind, heavy rainfall, and storm surge to the area along or close to the TCs' track. Among the three, heavy rainfall, which may lead to flash flooding and mudslides, is the most lethal natural disaster [1]. Typhoon Morakot, interacted with the strong southwest monsoon, produced copious amounts of rainfall in Taiwan, with a record of 3031.5 mm during August 6–13, 2009 [2]. Hurricane Katrina, one of the deadliest hurricanes in the history of the United States, brought over 15 inches (381 mm) heavy rainfall to Florida at its landfall. The recorded maximum sustained wind speed reached 175 mph (280 km/h) and wind gusts reached 220 mph (350 km/h) at New Orleans, Louisiana [3]. Typhoon Morakot left 619 people dead and the death toll due to hurricane Katrina was over 1100 [3–5]. In order to alleviate enormous loss of lives and properties in the future, it is important to notice the local population and civil authorities to make appropriate preparation for the cyclones, including evacuation of the vulnerable areas where necessary. Accurate and timely (24 and 12 h before landfall) forecasting the TC track and the potential rainfall and wind induced by TC are vital and essential [6].

Most operational meteorologists rely heavily on numerical weather prediction (NWP) models in forecasting TCs. TC track forecasts have improved significantly over the past several decades [1, 7–9]. By contrast, the improvement in forecasting the TCs intensity has lagged behind the progress of TCs' track forecasting [9–12]. The capability for NWP models to predict short-term rainfall is still very limited [1, 6, 13, 14]. Quantitative forecasting of rainfall remains problematic and lags behind the TC's track forecast, although tropical cyclone forecasting is a successful enterprise with favorable benefit-to-cost returns [1]. Kidder et al. [6] report that because few observations are available while the storm is offshore, initializing numerical weather prediction models with sufficient details of the storm is impossible. Therefore, the rainfall forecasts by NWP models are not so accurate. The research of Xu et al. [15] shows that currently in China, there is no effective operational approach to forecast the heavy rainfall and wind induced by tropical cyclone. The forecast of the rainfall and wind due to TC in China all relies on NWP models and the experience of forecasters. Exploring other ways to predict short-term rainfall is therefore important and necessary.

When TCs approach the land or move across the coast, the TCs structure and intensity change greatly [16]. Landfalling TCs usually bring about heavy rainfall over land. Regarding the forecasting of rainfall due to a TC in a certain region (at a certain rain gauge), it is reported that the rainfall is associated with the distance from the TC center, TC intensity, TC track, TC moving velocity, and TC residing time, as well as the environmental background. Rainfall induced by a TC generally decreases exponentially with distance from the TC center [6, 17, 18]. With the same environmental background, the stronger the TC intensity, the heavier the precipitation will be [18]. The distribution of precipitation due to landfalling TC is asymmetric. In the Northern Hemisphere, the land on the right-hand side of TC would usually receive more intense and spatial rainfall than the land on the left-hand side [19], since the rain bands on the right would carry more moist oceanic air than those on the left. After landfall, the slower the moving speed of the TC or the longer the residence time for the TC in a certain region, the more opportunities and longer time will be for the TC to interact with other weather systems, which might lead to extreme rainfall accumulation [2].

The previous studies indicate that rainfall induced by a TC at a certain rain gauge is attributable to a variety of factors. However, most of those studies focused on either case studies or investigating a specific factor, and the conclusion is mostly qualitative. In this study, a statistical

http://dx.doi.org/10.5772/64859

scheme will be developed to forecast the maximum daily rainfall and 3-day accumulative rainfall at a meteorological station by considering the factor of distance between the station and the TC-landfalling center, the intensity of the TC, and the landfalling direction of the TC. This paper is arranged as follows. An overview of the data and methodology is presented in Section 2. Section 3 contains the description of the statistical boxplot scheme for TCs rainfall. The applications of the boxplot scheme to forecast the rainfall due to TCs in 2012 and 2013 are described in Section 4. Summaries and conclusions are given in the final section.

2. Data and methodology

This study focuses on Shenzhen to explore the potential rainfall caused by landfalling TCs. Shenzhen is a coastal and urban city in Guangdong Province, China [20–23], with the latitude from 22°27′ to 22°52′ and longitude from 113°46′ to 114°37′ (**Figure 1**). In summer and autumn, the city is often influenced by TCs. In this study, a total of 427 TCs, which made landfall along the Southeast China coast from 1953 to 2011 with the landfalling distance within 700 km to Shenzhen meteorological station (SMS), are studied.

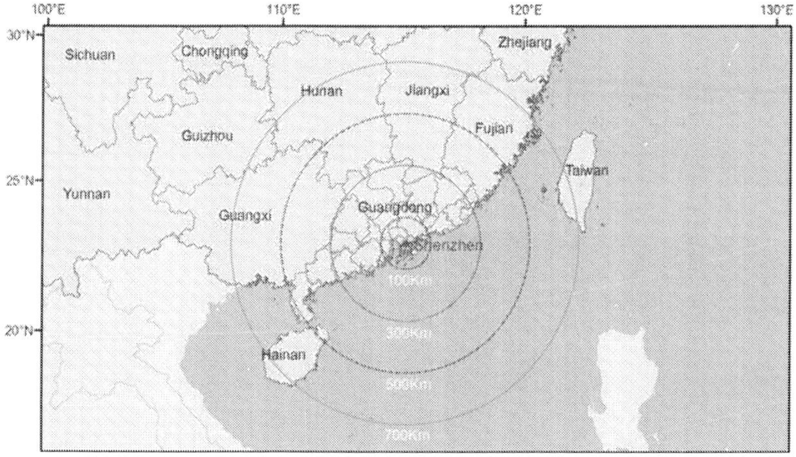

Figure 1. Location of SMS; circles indicate region with radii of 100, 300, 500, and 700 km to SMS.

The records of daily rainfall from 1953 to 2011 in SMS are obtained from the Shenzhen Meteorological Bureau (SZMB). The maximum daily rainfall (from 20 pm of the previous day to 20 pm of the current day based on China standard time) and the maximum 3-day accumulative rainfall at SMS during the TC-landfalling period (within a couple of days before or after landfall) are computed. For example, if a TC makes landfall on date A, the rainfall at SMS on date A−2, A−1, A, A+1, and A+2 is collected as R_{A-2}, R_{A-1}, R_A, R_{A+1}, and R_{A+2}. The maximum daily rainfall (R24) and the maximum 3-day accumulative rainfall (R72) are computed as follows:

$$R24 = \text{maximum}(R_{A-2}, R_{A-1}, R_A, R_{A+1}, R_{A+2}) \quad (1)$$

$$R72 = \text{maximum}(R_{A-2}R_{A-1}R_A, R_{A-1}R_AR_{A+1}, R_AR_{A+1}R_{A+2}) \quad (2)$$

where $R_{A-2}R_{A-1}R_A$ refers to the accumulative rainfall on date A−2, A−1, A, etc.

TC characteristics from 1953 to 2011 are collected from China Meteorological Administration (CMA). The TC characteristics include the landfalling track, the distance between the TC-landfalling center and SMS, and the intensity (maximum wind speed near the TC center), which are proved to be strongly related to the rainfall caused by TCs in a certain region [6, 17–19].

As the rainfall distribution and intensity on the right side of TC track is different from those on the left side of TC track [19], all the TCs are first grouped into two categories: A, TCs landfalling to the west of SMS (landfalling longitude <114°E); and B, TCs landfalling to the east of SMS (landfalling longitude >114°E). Next, A and B are further grouped into seven categories according to the landfalling distance to SMS, for example, A1, within 100 km; A2, 100–200 km; A3, 200–300 km; … A7, 600–700 km and B1, within 100 km; … B7, 600–700 km. Finally, A1, A2, …, B7 are grouped according to their landfalling intensity. According to the typhoon categorizing criterion of CMA (**Table 1**), there are six categories of TCs, which are super typhoon, severe typhoon, typhoon, severe tropical storm, tropical storm, and tropical depression. Among the 427 TCs, no landfalling TC was super typhoon. In this study, the TC intensity is stratified into three categories based on their respective TC-landfalling intensity scale (**Table 1**): TTY (total typhoon), which includes SuTY, STY, and TY; TTS (total tropical storm), which includes STS and TS; and TD. A1, A2, …, B7 are therefore grouped into A1-TTY, A1-TTS, A1-TD, A2-TTY, A2-TTS, A2-TD, …, B7-TTY, B7-TTS, and B7-TD. The flowchart of the TCs' categorizing steps is depicted in **Figure 2**.

Category	Abbreviation	Sustained maximum winds near the center of TCs
Super typhoon	SuTY	≥51 m/s
Severe typhoon	STY	41.5–50.9 m/s
Typhoon	TY	32.7–41.4 m/s
Severe tropical storm	STS	24.5–32.6 m/s
Tropical storm	TS	17.2–24.4 m/s
Tropical depression	TD	10.8–17.1 m/s

Table 1. Tropical Cyclone Intensity Scale according to CMA.

http://dx.doi.org/10.5772/64859

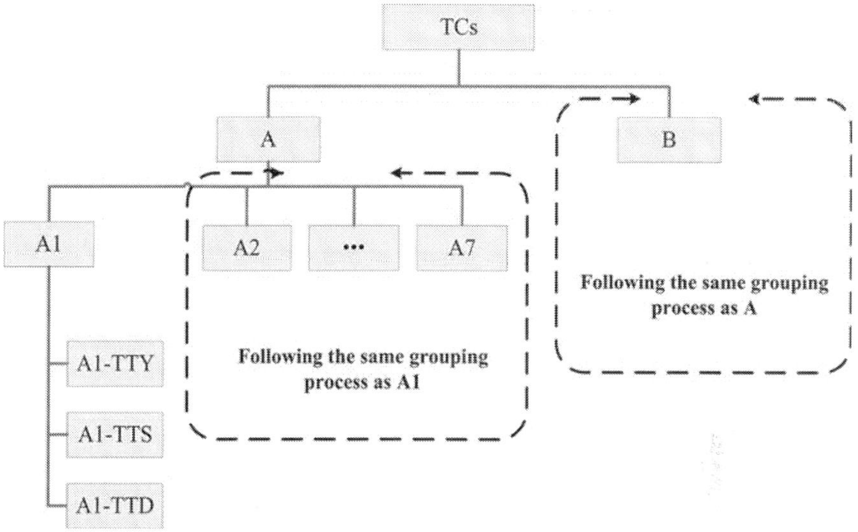

Figure 2. Flowchart of grouping process for landfalling tropical cyclone along the Southeast China coast.

Due to its nature of nonnormal distribution for precipitation [24], nonparametric method is a preferable approach to analyze typhoon-induced precipitation than parametric method [25]. A nonparametric statistical method, percentile estimation, is used to analyze the rainfall data in this study. Boxplots are applied to illustrate the analysis results. The detailed procedures are as follows:

Given n sorted rainfall observation $\{ri\}$, $i = 1, \ldots, n$, for a TC category, $0 < r_1 < r_2 < \ldots < r_n$, the corresponding percentile for a rainfall record, r_i, is computed to be [26]

$$P_i = \frac{100}{n}\left(i - \frac{1}{2}\right) \tag{3}$$

When computing the rainfall of any percentile P, it needs to find two consecutive p_k and p_{k+1}, where $p_k < P < p_{k+1}$. The rainfall for this p percentile is therefore

$$r = r_k + \frac{p - p_k}{p_{k+1} - p_k}(r_{k+1} - r_k) \tag{4}$$

Using this approach, the 25th, 50th, and 75th percentile rainfall for each TC category can be computed. The analysis results are illustrated by a boxplot (**Figure 3**) to display the differences between sample populations [27]. There are five-number summaries for a boxplot: the lower whisker (LW), lower quartile (Q1), median (Q2), upper quartile (Q3), and upper whisker (UW). Q1, Q2, and Q3 are the 25th, 50th, and 75th percentile of the population, respectively. Points are drawn as outliers if they are larger than Q3 + 1.5 × (Q3 - Q1) or smaller than Q1 − 1.5 × (Q3 − Q1). In **Figure 3**, two observations are considered as outliers, which are depicted by a red plus sign (+). The plotted whisker, UW (or LW) in the figure, is the maximum observation (or minimum observation), which is not an outlier.

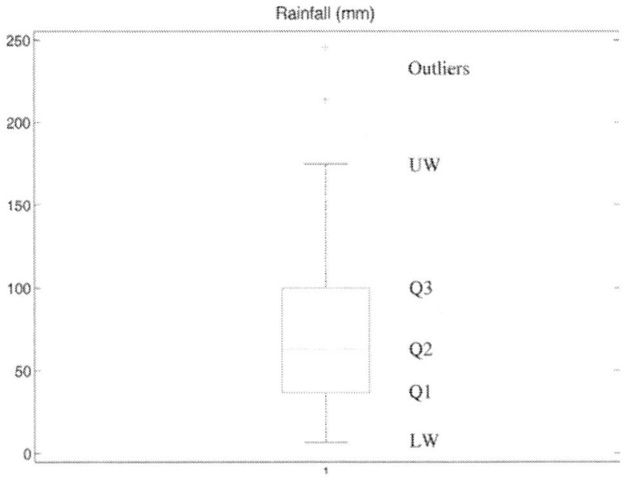

Figure 3. Boxplot as an example.

For practical use to forecast the rainfall due to a landfalling TC, the category of the landfalling TC is first determined based on the information of TC-landfalling distance to SMS from NWP models' forecast and the landfalling intensity from NWP models' forecast, as well as from the empirical experience of the forecasters. Then, referring to the boxplot corresponding to the landfalling TC category, the rainfall of R24 and R72 at the meteorological station can be estimated that the landfalling TC might cause the medium rainfall, 25–75% interquartile rainfall range, LW, UW, minimum rainfall, and maximum rainfall.

3. Statistical boxplots for TCs rainfall

Following the procedures mentioned in Section 2, all the historical TCs from 1953 to 2011, which made landfall within the distance of 700 km to SMS, are grouped into 42 categories. Because of the natural selection, it is impossible that each subgroup has approximately equal number of TCs. **Table 2** shows the groupings for TCs landfalling on the west of SMS, for example, "tms" in the table means the times of TCs for each category. For instance, there are 10 TTYs (2 severe typhoons and 8 typhoons) landfalling on the west of Shenzhen within 100 km distance to SMS (A1-TTY). The "tms" for other categories are from 4 to 29. For such small sample sizes, nonparametric statistics is seriously suggested to be suitable to analyze the datasets [25]. For each category of the TCs, the minimum, maximum, 25th percentile (Q1), 50th percentile (Q2), and 75th percentile (Q3) of the observed R24 and R72 are computed (**Table 2**).

A		R-24 (mm)					R-72 (mm)				
	tms	Min	Max	Q1	Q2	Q3	Min	Max	Q1	Q2	Q3
A1-TTY	10	93.7	247.4	120.3	152.1	226	102.1	347.9	185.8	263.1	290.8
A1-TTS	8	48.2	209.1	93.7	151.1	178.25	111.3	275.7	155	185.95	249.85
A1-TD	6	17.9	121.7	25.8	53.95	89.4	35.9	141.1	45.5	83.75	114.8
A2-TTY	4	101.1	213.5	114.5	140.1	182.9	138.2	478.5	154.1	235.3	389.55
A2-TTS	15	28.3	245.3	36.2	64.6	98.43	34.8	391.6	60.95	93.3	151.38
A2-TD	7	6.6	64.9	33.03	38	55.25	12.3	150.1	52.85	71.2	101.05
A3-TTY	6	23.5	80.9	41.1	60.9	80.3	35.4	131.8	58.5	98.35	114.3
A3-TTS	17	10.7	161	33.9	48.7	65.8	11.7	260	51.08	83.1	143.13
A3-TD	7	10.5	199.7	23.38	35.3	109.25	11.1	282.8	43.85	73.8	145.63
A4-TTY	6	41.9	73.4	54.7	62	68.2	42	128.2	71.3	84.4	90.4
A4-STS	15	4.1	156.6	7.73	36.5	45.95	5.8	261.2	14.7	45.1	96.78
A4-TD	5	15.9	102.5	18.98	22.4	48.2	16.6	105.8	34.08	40.1	65.38
A5-TTY	22	6.1	115	15.4	35.9	54.7	6.7	246.3	24.7	60.05	91.3
A5-TTS	29	0.6	168.8	11.2	26.9	47.5	0.8	189.6	16.58	43.5	74.13
A5-TD	21	1	102	9.1	35.1	46.95	2.4	157.6	13.73	47.4	93.23
A6-TTY	12	0.5	91.4	10.2	22.3	35.1	0.5	140.8	17.3	35.55	48.85
A6-TTS	21	1.4	159.2	6.45	14.7	28.83	1.5	282.5	9.28	24.2	58.73
A6-TD	21	1.1	308.6	6.35	19.2	56.2	1.2	387.8	8.9	31.6	85.78
A7-TTY	10	0.1	45.5	1.9	11.95	21.4	0.1	64.9	2.4	17.55	38
A7-TTS	10	1.2	64.5	6.8	25.85	53.3	1.3	110.1	13.7	52.95	73.7
A7-TD	11	0.1	159.2	1.23	10	33.35	0.1	282.5	2.13	18.3	74.55

Table 2. Times of each TC category and the corresponding percentile for R-24 and R-72 for TCs landfalling on the west of SMS.

The boxplots of the historical rainfall records of R24 and R72 at SMS due to all the categories of TCs from 1953 to 2011 are plotted in **Figure 4**. **Figure 4a** and **b** are for the category TTY, (c and d) are for TTS and (e and f) are for TD. The thick vertical line in each of the subplots of **Figure 4** refers to SMS. A7, A6, A5, A4, A3, A2, and A1 of the x-axis label refer to the TCs landfalling location, which are 600–700, 500–600, 400–500, 300–400, 200–300, 100–200 km, and within 100 km on the west of SMS. The meanings of B1, B2, ..., B7 are similar, but for the TCs landfalling on the east of SMS. Therefore, from **Figure 4**, the rainfall of R24 and R72 at SMS can be determined that each category of the landfalling TC might cause the 25–75% interquartile rainfall range, the lower whisker, upper whisker, minimum rainfall, and maximum rainfall.

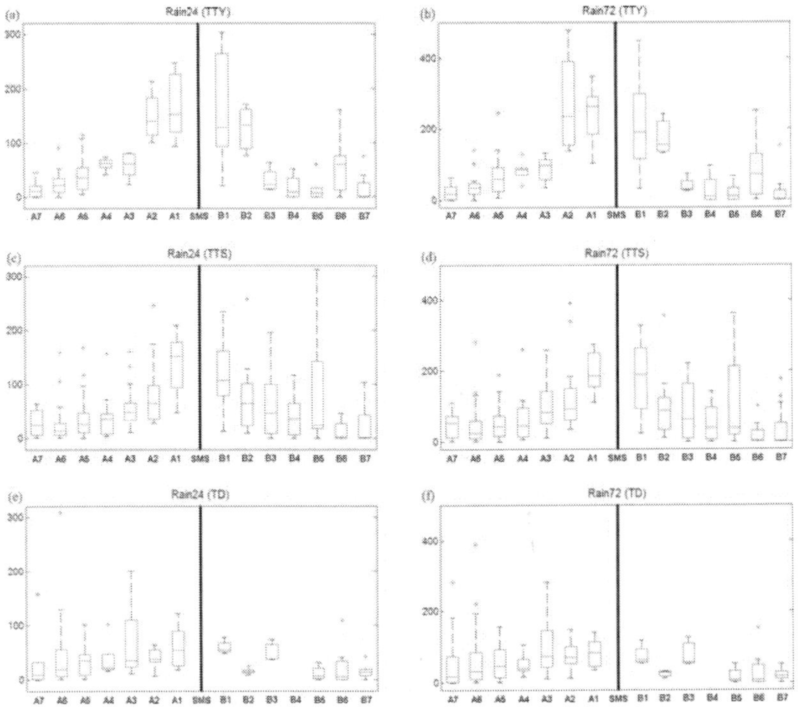

Figure 4. Boxplots for R24 and R72 rainfall at SMS from 1953 to 2011 due to TTYs (a and b), TTSs (c and d), and TDs (e and f). The vertical black thick line in the middle of each subplot denotes SMS. A on the *x*-label refers to the landfalling area, which is to the west of SMS, and B on the *x*-label refers to the landfalling area, which is to the east of SMS. The numbers after A and B on the *x*-label refer to the distance to SMS.

From **Figure 4**, it can be seen that the shape of the rainfall boxplots is not symmetric compared to the boxplots on the west and on the east of SMS. Generally, the TCs that make landfall on the west of SMS within the distance of 400 km would generally bring more rainfall to SMS than those TCs that make landfall on the east of SMS with similar landfalling intensity and distance. The median (Q2) records of R24 and R72 for TTY and TTS (**Figure 4a–d**) generally decrease with the increase of the landfalling distance to SMS, especially for TCs landfalling on the west of SMS. However, when the landfalling distance is outside 400 km, Q2 does not decrease as clearly as its change within 400 km. From **Figure 4a–d**, it can be seen that TCs that make landfall at B5 (with the distance of 400–500 km on the east of SMS) and B6 (with the distance of 500–600 km on the east of SMS) sometimes might cause very heavy rainfall at SMS. The reason might be that after those TCs land at B5-B6, they do not dissipate soon and continue to move west or east. During their continuous movement, they will probably interact with other weather system, such as southwesterly monsoon and mid-latitude trough, and bring plenty of moisture and energy over Southeast China. If SMS is on the passageway or near the passageway to transport such moisture and energy, heavy rainfall would occur in Shenzhen.

http://dx.doi.org/10.5772/64859

In addition from **Figure 4**, the variation of rainfall due to TDs does not change with distance as clearly as rainfall due to other categories of TCs. Furthermore, it can be seen from the figure that within the distance of 200 km, the median of the rainfall records (Q2) is generally larger when the strength of the TCs is higher (i.e., $Q2_{TTY}>Q2_{TTS}>Q2_{TD}$). However, when outside 200 km, this pattern might change. For example, rainfall at SMS due to TTSs and TDs, which make landfall at B3, is even higher than rainfall due to TTYs landfalling at B3. TTSs that make landfall at B5 (more than 400 km away from SMS) might sometimes induce very large rainfall at SMS (**Figure 4c, d**), compared to TTYs that make landfall at the B5. This might be due to the influence of other factors, such as TC track, TC-moving velocity after landfalling, and the environmental background.

4. 2012, 2013 cases: forecasting and discussion

To apply the statistical boxplot scheme to forecast the potential maximum R24 and R72 rainfall due to a landfalling TC, the TC-landfalling location and intensity need to be identified first. Nowadays, operational forecasting models can usually predict the TC's track well, especially the track within 24 h [15]. Compared to the track forecast, the TC intensity forecast by NWP models is not so accurate at present. However, with the real-time observations from satellite and radar, forecasters have the ability to predict the intensity of the landfalling TC 12 h before TC landfall by rule of thumb, as it usually takes at least 12 h for a TC to change intensity or motion appreciably [1]. With the approximate TC-landfalling location and intensity available 12 h before the TC landfall, the statistical boxplot scheme can be then applied to forecast the rainfall of R24 and R72 at SMS.

In this section, the TCs that made landfall along the Southeast China coast in 2012 and 2013 will be used to test the performance of the boxplot scheme. In the typhoon seasons of these 2 years, there are 11 TCs landfalling within the distance of 700 km to SMS along the Southeast China coast. The detailed TC information is summarized in **Table 3**.

Based on the latest forecast for these 11 TCs' landfalling direction, location, and intensity, the boxplots for the corresponding TC categories are picked out. **Figure 5** shows these boxplots for the rainfall of R24, as well as the real maximum daily rainfall observation at SMS (black thick solid horizontal line) during the TC-landfalling period (rainfall within a couple of days before and after TC landfalling) due to these TCs. By comparison, the latest rainfall forecasts (brown solid horizontal line) before the TCs' landfall by European Center for Medium-Range Weather Forecasts (ECMWF) are plotted in **Figure 5** as well. ECMWF is renowned worldwide for providing the most accurate medium-range global weather forecasts up to 10 days ahead, monthly forecasts, and seasonal outlooks to 6 months ahead. It has been widely used for operational forecast and research purpose around the world [28–31]. ECMWF model is reported to well produce the medium-range forecasts of the Northwest Pacific subtropical high and South Asian high which have pronounced influences on the summertime persistent heavy rainfall in China [32]. The track of the TC in western Pacific and South China Sea is strongly affected by the area of the western Pacific subtropical high [33, 34]. Therefore, ECMWF's forecasts of TC's track for the coming 12 and 24 h are usually reliable. SZMB has been relying on ECMWF for daily operational forecast since 2009. The finest resolution of ECMWF system used in SZMB is 0.125° × 0.125°. Xu et al. [28] reported that ECMWF model could provide valuable information for rainfall forecast up to the next 10 days; however, the ability for storm rainfall forecast was not reliable. For tropical cyclone forecast, ECMWF can usually accurately predict the TC's landfalling intensity and location by the latest forecast around 12 h before TC's landfall, but it cannot predict well the storm rainfall induced by TCs.

Name	Landfalling time	Landfalling latitude (°N)	Landfalling longitude (°E)	Landfalling intensity	Distance (km)
Doksuri	6/30/03	22	113.2	TTS	102
Vicente	7/24/04	22	113	TTY	119
Kai-Tak	8/17/12	21	110.4	TTY	410
TD1303	6/15/17	19.9	110.9	TD	436
Bebinca	6/22/11	19.2	110.7	TTS	506
Rumbia	7/02/06	21.1	110.2	TTS	424
Cimaron	7/18/22	24.1	117.9	TTS	434
Jebi	8/02/20	19.7	110.9	TTS	451
Utor	8/14/15	21.6	111.9	TTY	241
Trami	8/22/03	25.7	119.5	TTY	659
Usagi	9/22/20	22.7	115.4	TTY	263

Time format for landfalling time: month/day/hour; distance refers to the distance between the landfalling center of TC and SMS. The unit for latitude is degree north (°N) and for longitude is degree east (°E). The unit for distance is kilometer (km).

Table 3. Information about the 11 TCs landfalling within the distance of 700 km to SMS in 2012 (the first three TCs) and 2013 (the last eight TCs).

From **Figure 5**, it can be seen that the observed R24 rainfall records at SZMB for the 11 TCs are most of the time between the historical observed maximum and minimum rainfall for the corresponding category of landfalling TC, except for TY Trami and STY Usagi. TY Trami landed in Fujian province, China on August 22, 2013. It continued to move west to Jiangxi Province, until it finally disappeared in Hunan Province. During its process in mainland, Trami interacted with the strong southwest monsoon and brought immense downpours in Southeast China. It set a new rainfall record at Shenzhen for TTY category, which landed on the east of Shenzhen with the landfalling distance of 600–700 km. Similar condition is for STY Usagi. Usagi set a new rainfall record at Shenzhen for TTY category, which landed on the east of Shenzhen with the landfalling distance of 200–300 km. For the other TCs, most of the rainfall observations at SMS are within the interquartile range of the historical records (Q1–Q3), such as TS Doksuri, TY Vicente, TY Kai-Tak, TD1303, STS Rumbia, STS Jebi, and STY Utor, which are shown in

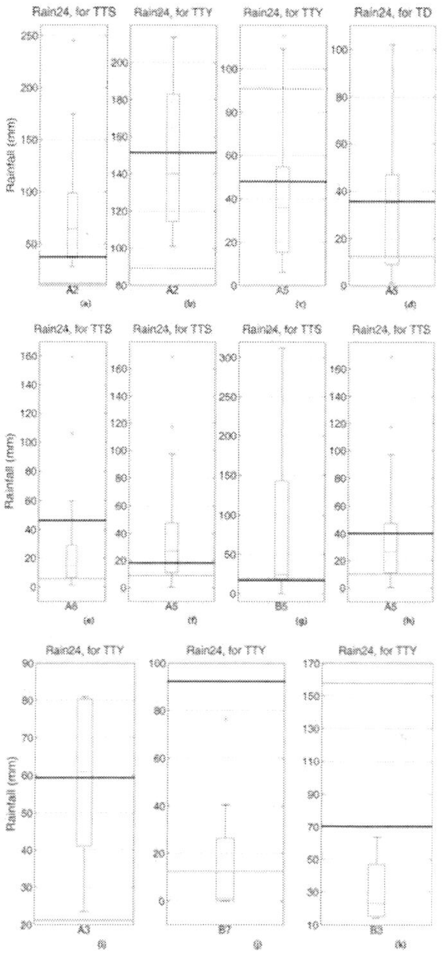

Figure 5. Comparison of the R24 rainfall forecasts for 11 TC cases in 2012 and 2013 by the boxplots and by ECMWF (brown solid horizontal line), as well as the real rainfall observation (black thick solid horizontal line): (a) for TS Doksuri; (b) for TY Vicente; (c) for TY Kai-Tak; (d) for TD1303; (e) for TS Bebinca; (f) for STS Rumbia; (g) for STS Cimaron; (h) for STS Jebi; (i) for STY Utor; (j) for TY Trami; and (k) for STY Usagi.

Figure 5a–d,f,h,i, respectively. Similarly, more than 50% of R72 rainfall at SMS due to these TCs are within the interquartile range (Q1–Q3) of the historical records. By comparison, it can be seen that ECMWF by the latest model output can sometime accurately forecast the rainfall due to TCs, such as in **Figure 5f** and **g**. However, for the other cases, the discrepancies between the rainfall forecast by ECMWF and real rainfall observation at SMS are large. Therefore, besides the NWP, forecasters can use the historical rainfall observation boxplots (**Figure 4**) as a good reference to predict the potential short-term rainfall due to a landfalling TC.

The statistical boxplots scheme can provide valuable information to operational forecaster to predict the potential rainfall due to a TC. However, it must be known that there are still many uncertainties for the boxplots because of the small sample size for some TC categories due to the natural features, as well as the short observation history. With more TCs landfalling along the Southeast China coast in future, the larger will be the database of TCs and the more accurate will be for the short-term TC rainfall prediction by the boxplots scheme.

5. Conclusions

This study applied all the historical TCs landfalling along the Southeast China coast from 1953 to 2011 to explore a statistical boxplot scheme to forecast the maximum daily (R24) and 3-day accumulative rainfall (R72) at a certain rain gauge (SMS) during TC-landfalling period. Three TC's characteristics, including TC's landfalling direction, landfalling distance to SMS, and landfalling intensity, are considered to categorize all the historical TCs landfalling within the distance of 700 km to SMS. The corresponding historical daily rainfall records at SMS during the TC-landfalling period (rainfall within a couple of days before and after TC landfalling) are collected and organized according to the TC's category. The organized rainfall records for each TC category are analyzed by percentile estimation. The results are plotted in boxplots. It is concluded from the boxplots that the rainfall at a certain area is generally positively correlated to the intensity of the landfalling TC within 200-km distance to the landfalling center. Within the distance of 400 km to the landfalling location, the rainfall at SMS is generally negatively associated with the distance between TC's landfalling center and SMS. With the same intensity scale, TCs landfalling on the west of SMS will generally bring heavier rainfall at SMS than TCs landfalling on the east of SMS within the distance of 400 km. Eleven tropical cyclones landfalling within the distance of 700 km to SMS in 2012 and 2013 are used to evaluate the performance of the statistical boxplots to predict short-term rainfall at SMS. Results show that the boxplot scheme is very valuable to forecasters to provide rainfall range due to each TC category. For most of the time, the observed rainfall due to a landfalling TC is within the range of the historical rainfall records.

The boxplot scheme is easy to implement. The rainfall boxplots are quite helpful to operational forecasters. As of this writing, the technique is already in use as a valuable reference at SZMB to predict the short-term rainfall at SMS due to TCs.

Acknowledgements

This paper is supported by the Natural Science Foundation of Guangdong Province with Grant 2015A030313742, Special Fund for Science and Technology Development in Guangdong Province with Grant No. 2016A050503035, and The Innovation of Science and Technology Commission of Shenzhen Municipality with Grants JCYJ20120617115926138 and ZDSYS20140715153957030.

http://dx.doi.org/10.5772/64859

Author details

Qinglan Li[1*], Hongping Lan[2], Johnny C.L. Chan[3], Chunyan Cao[2], Cheng Li[2] and Xingbao Wang[1,4]

*Address all correspondence to: ql.li@siat.ac.cn

1 Shenzhen Institute of Advanced Technology, Chinese Academy of Sciences, Shenzhen, China

2 Shenzhen Meteorological Bureau, Shenzhen, China

3 School of Energy and Environment, City University of Hong Kong, Hong Kong, China

4 The Centre for Australian Weather and Climate Research, Bureau of Meteorology, Melbourne, VIC, Australia

References

[1] Willoughby H.E., Rappaport E.N., Marks F.D. Hurricane Forecasting: The State of the Art. Natural Hazards Review. 2007; 8(3): 45–49.

[2] Wu L., Liang J., Wu C. Monsoonal Influence on Typhoon Morakot (2009). Part I: Observational Analysis. Journal of the Atmospheric Sciences. 2011; 68: 2208–2221. DOI: http://dx.doi.org/10.1175/2011JAS3730.1.

[3] Jonkman S.N., Maaskant B., Boyd E., Levitan M.L. Loss of life caused by the flooding of New Orleans after Hurricane Katrina: analysis of the relationship between flood characteristics and mortality. Risk Analysis: An Official Publication of the Society for Risk Analysis. 2009; 29(5): 678–98. DOI: 10.1111/j.1539-6924.2008.01190.x.

[4] Cui P., Chen S., SU F., Zhang J. Formation and Mitigation Countermeasure of Geo-Hazards Caused by Moarc Typhoon in Taiwan. Journal of Mountain Science. 2010; 28(1): 103–115.

[5] Knabb R.D, Rhome J.R., Brown D.P. National Hurricane Center. Tropical Cyclone Report: Hurricane Katrina, 23–30 August 2005 [Internet]. 20 December 2005 [Updated: 14 September 2011]. Available from: http://www.nhc.noaa.gov/data/tcr/AL122005_Katrina.pdf [Accessed: November 1, 2013].

[6] Kidder S.Q., Kusselson S.J., Knaff J.A, Ferraro R.R, Kuligowski R.J., Turk M. The Tropical Rainfall Potential (TRaP) Technique. Part I: Description and Examples. Weather and Forecasting. 2005; 20(4): 456–464.

[7] Aberson S.D. The Ensemble of Tropical Cyclone Track Forecasting Models in the North Atlantic Basin (1976–2000). Bulletin of the American Meteorological Society. 2001; 82(9): 1895–1904.

[8] Franklin J.L., Mcadie C.J., Lawrence M.B. Trends in Track Forecasting for Tropical Cyclones Threatening the United States. Bulletin of the American Meteorological Society. 2003; 84(9): 1197–1203.

[9] Rogers R., Aberson S., Black M., Black P., Cione J., Dodge P., et al. The Intensity Forecasting Experiment: A NOAA Multiyear Field Program for Improving Tropical Cyclone Intensity Forecasts. Bulletin of the American Meteorological Society. 2006; 87(11): 1523–1537.

[10] DeMaria M., Gross J.M. Evolution of Tropical Cyclone Forecast Models. In: Simpson R., editor. Hurricane! Coping with Disaster. 1st ed. American Geophysical Union; 2003. p. 360. DOI: ISBN 0-87590-297-9.

[11] Cecil D.J., Jones T.A., Knaff J.A., DeMaria M. Statistical Forecasting of Pacific and Indian Ocean Tropical Cyclone Intensity using 19-, 37-, and 85- GHZ Brightness Temperatures. In: 26th Conference on Hurricanes and Tropical Meteorology; May 5, 2004; Miami, FL. American Meteorological Society; 2004. p. 302–303.

[12] Knaff J.A., Sampson C.R., DeMaria M. An Operational Statistical Typhoon Intensity Prediction Scheme for the Western North Pacific. Weather and Forecasting. 2005; 20(4): 688–699.

[13] Marchok T., Rogers R., Tuleya R. Validation Schemes for Tropical Cyclone Quantitative Precipitation Forecasts: Evaluation of Operational Models for U.S. Landfalling Cases. Weather and Forecasting. 2007; 22(4): 726–746.

[14] Liu G., Chao C., Ho C. Applying Satellite-Estimated Storm Rotation Speed to Improve Typhoon Rainfall Potential Technique. Weather and Forecasting. 2008; 23(2): 259–269.

[15] Xu Y.L., Zhang L., Gao S.Z. The Advances and Discussions on China Operational Typhoon Forecasting (in Chinese). Meteorological Monthly. 2010; 36(7): 43–49.

[16] Chen L.S. Research Progress on the Structure and Intensity Change for the Landfalling Tropical Cyclones. Journal of Tropical Meteorology. 2012; 18(2): 113–118.

[17] Simpson R.H., Riehl H. The Hurricane and its Impact. Louisiana State University Press, Baton Rouge; 1981. 398 p.

[18] Pfost R.L. Operational Tropical Cyclone Quantitative Precipitation Forecasting. National Weather Digest. 2000; 24(1–2): 61–66.

[19] Chan J.C.L., Liang X. Convective Asymmetries Associated with Tropical Cyclone Landfall. Part I: f -Plane Simulation. Journal of the Atmospheric Sciences. 2003; 60(13): 1560–1576.

[20] Wang M.J., Zhang X.L., Li X.R. Analysis of Meteorological Conditions for the Base of Marine Sports in the 26th Summer Universiade in Shenzhen in 2011. Journal of Tropical Meteorology. 2011; 17(2): 187–192.

[21] Zhang X.L., Li L., Du Y., Jiang Y., Fang X.Y., Li M., et al. A Numerical Study on the Influences of Urban Planning and Construction on the Summer Urban Heat Island in the Metropolis of Shenzhen. Journal of Tropical Meteorology. 2011; 17(4): 392–398.

[22] Chen J., Li Q.L., Niu J., Sun L.Q. Regional Climate Change and Local Urbanization Effects on Weather Variables in Southeast China. Stochastic Environmental Research and Risk Assessment. 2011; 25(4): 555–565. DOI: 10.1007/s00477.

[23] Li Q.L., Chen J. Teleconnection between ENSO and climate in South China. Stochastic Environmental Research and Risk Assessment. 2014; 28(4): 927–941. DOI: 10.1007/s00477-013-0793-z.

[24] Turco M., Llasat M.C. Trends in Indices of Daily Precipitation Extreme in Catalonia (NE Spain), 1951-2003. Natural Hazards and Earth System Sciences. 2011; 11: 3213–3226.

[25] Pett M.A. Nonparametric Statistics for Health Care Research – Statistics for Small Samples and Unusual Distributions. 1st ed. Thousand Oaks: Sage Publications; 1997. 307 p.

[26] Stark H., Woods J. Probability, Statistics, and Random Processes for Engineers. 4th ed. Boston: Pearson; 2012. 704 p.

[27] McGill R., Tukey J.W., Larsen W.A. Variations of Box Plots. Journal of the American Statistical Association. 1978; 32(1): 12–16.

[28] Xu W.W., Chen S.P., Li Q.L. Evaluation of Precipitation Forecast in Shenzhen for the First Raining Season in 2012 by ECMWF model and HAPS model (in Chinese). Guangdong Meteorology. 2013; 35(5): 6–9.

[29] Halperin D.J., Fuelberg H.E., Hart R.E., Cossuth J.H., Sura P. An Evaluation of Tropical Cyclone Genesis Forecasts from Global Numerical Models. Weather and Forecasting. 2013; 28(6): 1423–1445. DOI: http://dx.doi.org/10.1175/WAF-D-13-00008.1.

[30] Wang Y., Bellus M., Geleyn J.F., Ma X.L., Tian W.H., Weidle F. A New Method for Generating Initial Condition Perturbations in a Regional Ensemble Prediction System: Blending. Monthly Weather Review. 2014; 142(5): 2043–2059.

[31] Magnusson L., Bidlot J.R., Lang S.T.K., Thorpe A., Wedi N. Evaluation of Medium-Range Forecasts for Hurricane Sandy. Monthly Weather Review. 2014; 142(5): 1962–1981.

[32] Niu R.Y., Zhai P.M. Synoptic Verification of Medium-Extended-Range Forecasts of the Northwest Pacific Subtropical High and South Asian High Based on Multi-Center TIGGE Data. Acta Meteorologica Sinica. 2013; 27(5): 725–741.

[33] Sun Y., Zhong Z., Lu W., Hu Y.J. Why Are Tropical Cyclone Tracks over the Western North Pacific Sensitive to the Cumulus Parameterization Scheme in Regional Climate Modeling? A Case Study for Megi (2010). Monthly Weather Review. 2014; 142(3): 1240–1249.

[34] Tao L., Li S.J. Impact of Tropical Intraseasonal Oscillation on the Tracks of Tropical Cyclones in the Western North Pacific. Journal of Tropical Meteorology. 2014; 20(1): 26–34.

Chapter 11

Climate Risk Early Warning System for Island Nations: Tropical Cyclones

Yuriy Kuleshov

Additional information is available at the end of the chapter

http://dx.doi.org/10.5772/64029

Abstract

Tropical cyclones (TCs) frequently affect coastal areas of Australia and islands in the tropical Indian and Pacific oceans. Multi-hazards associated with TCs (destructive winds, storm surges and torrential rain) have dramatic impact on population and infrastructure. Accurate forecasting of TC seasonal activity is an important part of a Climate Risk Early Warning System (CREWS) for improving resilience of the society to potentially destructive impacts of TCs. Currently, a statistical model-based prediction of TC activity in the coming season is used for operational seasonal forecasting in the Australian region and the South Pacific Ocean. In this chapter, a possibility of improving the accuracy of seasonal TC prediction using advanced statistical model-based approaches is demonstrated. It is also demonstrated that an alternative approach—dynamical (physics-based) climate modelling—is promising for skilful seasonal TC forecasting. Using improved statistical and dynamical model-based methodologies for TC seasonal prediction as an integral part of the CREWS will provide valuable information about TC seasonal variability and will assist with decision making, responses and adaptation in island countries.

Keywords: tropical cyclones, early warning system, multi-hazards, preparedness

1. Introduction

Tropical cyclones (TCs) frequently affect coastal communities of Australia and island nations in the Indian and Pacific oceans, and pose significant threat to life and property. In many cases TC impacts on island countries were devastating [1]. Knowledge about TC variability (spatial and temporal) is important for improving preparedness and resource mobilisation well in advance of potential TC impacts, to reduce risk of multi-hazards associated with TCs.

TC activity is affected by climate variability and climate change. Large-scale climatic modes such as the El Niño-Southern Oscillation (ENSO) are known to modulate TC occurrences [2]. In addition, modelling of future climate suggests likely changes in TC activity. As stated in Chapter 14, Climate Phenomena and their Relevance for Future Climate Change of the IPCC Fifth assessment report: 'Based on process understanding and agreement in twenty-first century projections, it is *likely* that the global frequency of occurrence of tropical cyclones will either decrease or remain essentially unchanged, concurrent with a *likely* increase in both global mean tropical cyclone maximum wind speed and precipitation rates' [3].

To improve our knowledge about historical TCs in the Indian and Pacific oceans and develop accurate methodologies for TC seasonal forecasting, 'Climate Change and Southern Hemisphere Tropical Cyclones' International Initiative has been established in 1999 [4]. This International Initiative addressed three key areas: (i) preparing high quality TC historical database, (ii) producing TC climatology and (iii) developing skilful methodologies for TC seasonal prediction.

TC seasonal forecasting is one of the important elements of a Climate Risk Early Warning System (CREWS) aiming to increase preparedness of coastal communities at risk. Over the last few decades, statistical model-based methods for TC seasonal forecasting have been developed, starting with the pioneering work of Gray [5]. Statistical models are based on historical relationships of TC activity with large-scale environmental drivers which modulate TC activity such as the ENSO. Relating the observed numbers of TCs with ENSO indices it is possible to derive linear regression equations which can be used for prediction of future cyclone activity. The TC-ENSO relationship was used in developing statistical methodology for forecasting seasonal TC activity in the Australian and some other regions [6–8].

However, in a globally warming environment, statistical models may not produce reliable outcomes when values of ENSO indices are outside of the range of historical records. While the developed statistical models performed reasonably well in the past, during a very strong La Niña event in 2010–2011 the statistical models significantly over-predicted the number of TCs in the Australian region [9]. It became evident that improving statistical methodologies and developing new dynamical climate model-based methodologies is essential to improve prediction skill.

In this chapter, prospects for improving the skill of operational seasonal prediction of TC activity in the regions of the Southern Hemisphere (SH) using statistical and dynamical model-based approaches are presented.

2. Southern Hemisphere tropical cyclone archive and data portal

2.1. Tropical cyclone historical data archive

Accurate historical cyclone records (preferably long-term records covering a few decades) are required for reliable prediction of future TC activity. Thus, the first objective of the 'Climate Change and Southern Hemisphere Tropical Cyclones' International Initiative was to prepare

http://dx.doi.org/10.5772/64029

a high quality historical database of occurrences of TCs in the Indian and Pacific oceans. Historical TC records have been significantly improved since the 1970s due to availability of satellite imagery [10–12] and they were extensively used for preparing the Southern Hemisphere (SH) TC archive.

The SH TC archive has been prepared at the National Climate Centre (NCC) of the Australian Bureau of Meteorology during 1999–2003 in collaboration with the National Meteorological and Hydrological Services (NMHSs) of Fiji, France and New Zealand (NZ). The first version of the SH TC archive has been released in 2003 [13]. Since then, historical data are regularly updated to keep the archive up to date.

Updating the archive is a two-step procedure which includes (i) collection of best track data (or operational data if best track data are not available) from Tropical Cyclone Warning Centres (TCWCs) in Brisbane, Darwin and Perth (Australia), Jakarta (Indonesia), Port Moresby (PNG), Wellington (NZ), Regional Specialised Meteorological Centres (RSMCs) La Reunion (France) and Nadi (Fiji) and (ii) combining the data in one consolidated archive including quality control, correction for errors and making a consensus expert decision when joining tracks of systems which occurred in two or three areas of responsibilities of different TCWCs and RSMCs.

Recently, as a part of the Pacific Australia Climate Change Science and Adaptation Planning (PACCSAP) program's 'Seasonal tropical cyclone prediction' project, the SH TC archive has been revised. Specifically, data for 1969–1970 to 2010–2011 TC seasons covering the South Pacific Ocean and produced by RSMC Nadi (Fiji), TCWCs Brisbane and Darwin (Australia) and TCWC Wellington (New Zealand) have been examined to eliminate errors and inconsistencies. The following rules have been applied when preparing a consolidated archive. As RSMC Nadi (Fiji) is a designated by the World Meteorological Organization (WMO) centre with responsibilities to issue TC warnings and prepare best track data for the area between the equator and 25°S, 160°E and 120°W, its data have been treated as a primary source of information for this area. However, RSMC Nadi was established in 1995 while the SH TC archive extends to cover TC seasons from the 1970s (satellite era). Thus, TC best track data prepared for this area by TCWCs in Brisbane, Darwin and Wellington have been used for 1969–1970 to 1994–1995 TC seasons, and data from RSMC Nadi—from the 1995–1996 TC season onwards. As for the other areas of the South Pacific Ocean, TC best track data from TCWCs in Brisbane and Darwin for the Australian region (between the equator and 37°S, 135°E and 160°E) and from TCWC Wellington for the New Zealand region (between 25°S and 40°S, 160°E and 120°W) have been used for entire length of records from 1969–1970 to 2010–2011 TC seasons.

As a result of growth of the 'Climate Change and Southern Hemisphere Tropical Cyclones' International Initiative and its geographic expansion to cover the Western North Pacific region, TC best track data produced by RSMC Tokyo for 1977–2011 seasons have been added to the consolidated archive. Similarly, TC best track data produced by RSMC la Réunion for 1969–2011 have been added to cover the South Indian Ocean region.

2.2. Tropical cyclone data portal

Tropical cyclone data portal has been created with aims (i) to visualise the data from the SH TC archive and (ii) allow users to perform analysis of historical cyclone data. Based on recent changes of and additions to the SH TC archive, the TC historical data portal has been re-designed to incorporate best track data for the Western Pacific both south and north of the equator and the South Indian Ocean (**Figure 1**).

Figure 1. Front page of the portal – the Southern Hemisphere (top panel) and the Western North Pacific (bottom panel).

New functionality has been also added to the portal including enhanced spatial and temporal selection of cyclones. Some examples of the portal's new functionality are given below. The portals are extensively used by NMHSs of island countries in the Pacific and Indian oceans for analysis of historical TC data and consequently it is reflected in the examples.

Analysis of historical TC tracks is often required to examine an individual cyclone's impact on a specific location. Such analysis could be performed using 'Place name' and 'Coordinates' options of the portal (**Figures 2** and **3**, respectively).

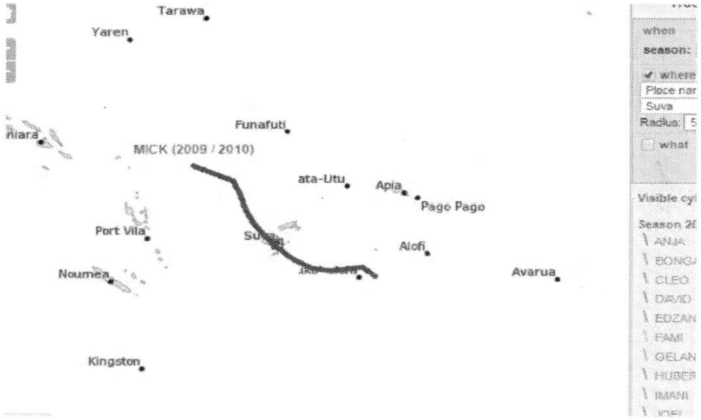

Figure 2. Track of TC *Mick* affecting Suva, Fiji displayed after selecting 'Place name' option.

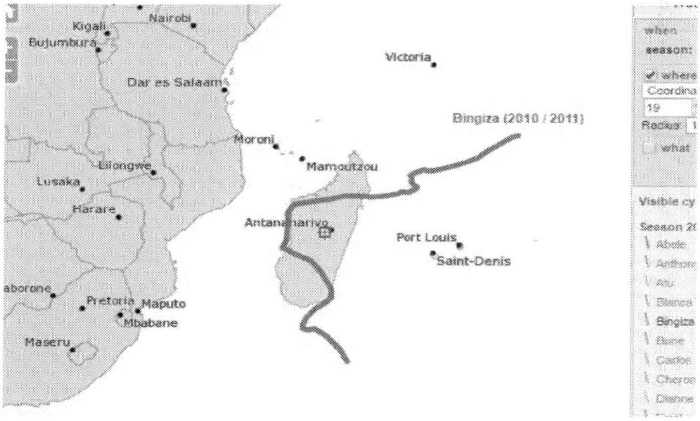

Figure 3. Track of TC *Bingiza* affecting area within 100 km radius of Antananarivo, Madagascar displayed after selecting 'Coordinates' option.

Similarly, analysis of historical TC tracks is often required to examine occurrences of TCs over larger areas, e.g. an exclusive economic zone of an island country, during a specific season or a number of seasons (**Figures 4** and **5**, respectively).

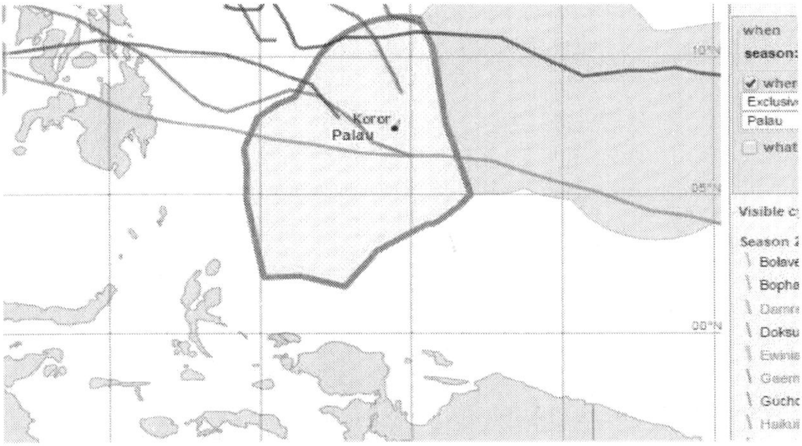

Figure 4. Tracks of TCs which passed through an exclusive economic zone of Palau during 2012 season.

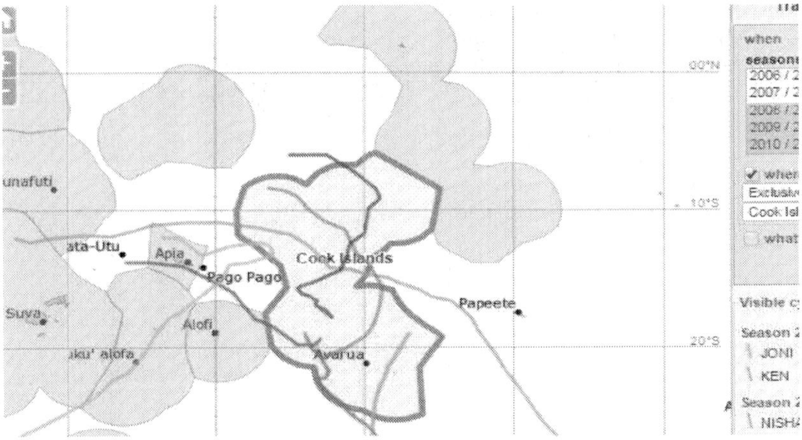

Figure 5. Track of TCs which passed through an exclusive economic zone of Cook Islands from 2008/2009 to 2010/2011 seasons.

Information about cyclone's occurrence (**Figure 6**) and changes of its intensity (**Figure 7**) is also incorporated in functionality of the portal.

http://dx.doi.org/10.5772/64029

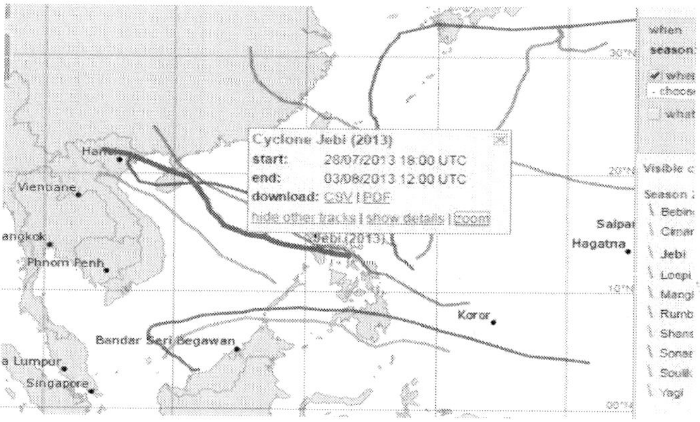

Figure 6. Tracks of TCs over selected area in the Western North Pacific in 2013 season together with information about start and end of TC *Jebi*.

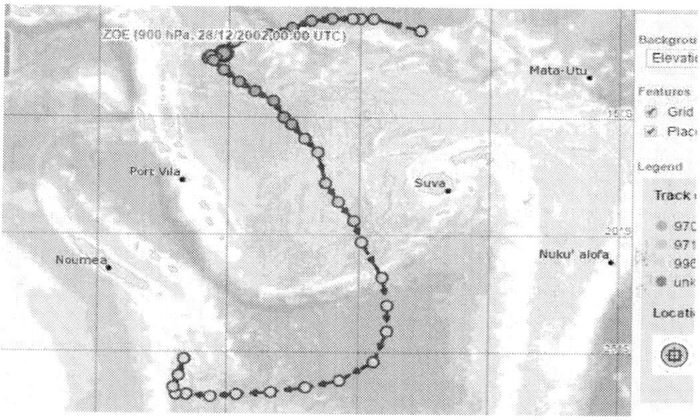

Figure 7. Track of TC *Zoe* displayed over 'Elevation and Bathymetry' background. Changes in the cyclone's intensity are colour-coded; legend is displayed on the right side.

3. Statistical prediction models

Currently, a statistical technique is used by the Australian Bureau of Meteorology to prepare operational TC seasonal prediction for the Australian and the South Pacific Ocean regions. The operational statistical model consists of linear discriminant analysis (LDA) models, based on ENSO indices as predictors.

3.1. The ENSO indices and linear discriminant analysis model

The Southern Oscillation Index (SOI) and sea surface temperature anomalies (SSTAs) in Niño3.4 and Niño4 regions (NIÑO3.4 and NIÑO4) are commonly used indices in defining ENSO phases which describe oceanic and atmospheric responses, respectively.

It has been demonstrated by Kuleshov et al. [2, 7] and Ramsay et al. [14] that in the Australian region a strong correlation (about −0.7) exists between the annual number of TCs and the NIÑO4 and NIÑO3.4 indices averaged over 3 months preceding the onset of a Southern Hemisphere TC season (August-September-October). In the eastern South Pacific Ocean, better correlation of the annual TC number with ENSO indices was found for the NIÑO3.4 and the SOI [7]. Based on these findings, the NIÑO3.4 and the SOI indices have been selected by the Australian Bureau of Meteorology for use in operational LDA statistical TC-ENSO model for seasonal prediction of TCs in both the Australian and the South Pacific regions.

A multivariate ENSO index has been developed at the NCC with the aim to integrate both atmospheric and oceanic responses in one index [2]. It is based on the first principal component of monthly Darwin mean sea level pressure (MSLP), Tahiti MSLP and the NIÑO3, NIÑO3.4 and NIÑO4 SST indices [2, 7]. This standardised monthly anomaly index is usually denoted as the 5VAR index. Further examining correlation of ENSO indices with TC occurrences in the Australian region, Kuleshov et al. [15] found that the 5VAR performs better than the SOI and NIÑO3.4 demonstrating the strongest monthly (−0.67, pre-season September), bi-monthly (−0.67, August and September) and tri-monthly correlation (−0.66, July, August and September).

Incorporating into statistical model a decreasing trend in TC activity over the Australian region in recent years [2, 16] and using 5VAR, SOI and NIÑO3.4 indices as predictors, Kuleshov et al. [15] demonstrated potential for improving skill of the LDA operational model. Brief description of the developed statistical model is presented in the Appendix. Cross-validation employed to assess the models' performance demonstrated that the models which used the pre-season July-August-September SOI and September 5VAR indices and the time trend as the predictors [15] demonstrated increased skill in TC seasonal forecasting compared with currently used LDA model [7].

3.2. Support vector regression (SVR) models

Recently, application of advanced statistical methodologies for seasonal prediction of TCs has been explored. It has been demonstrated that improvement in prediction skill compared to the LDA model can be achieved using support vector regression (SVR) models, exploring new environmental indicators and non-linear relationships between them [9]. Detailed description of the developed SVR models for the Australian and South Pacific Ocean regions could be found in [17] and its brief description is presented in the Appendix. Analysis of the results of the SVR models shows that the Dipole Mode Index, the 5VAR index and the SOI are the most frequently used indices selected for TC seasonal forecasting in the Australian and South Pacific regions.

http://dx.doi.org/10.5772/64029

4. Dynamical climate models

Dynamical climate modelling is an alternative to statistical modelling. Early analysis has revealed that the dynamical seasonal prediction system Predictive Ocean Atmosphere Model for Australia (POAMA) has skill in the prediction of ENSO which modulates TC activity in the SH [2].

As a part of the 'Climate Change and Southern Hemisphere Tropical Cyclones' International Initiative, an evaluation of performance of dynamical climate models for TC seasonal prediction was conducted under the PACCSAP program. The Australian Bureau of Meteorology and the Japan Meteorological Agency/Meteorological Research Institute (JMA/MRI) have developed systems to provide predictions of TC activity based on their dynamical models.

The two agencies each have their own coupled seasonal forecast model comprised of a number of ensemble members and a 30-year hindcast period. The JMA/MRI-CGCM is used by the JMA/MRI; the Bureau uses POAMA. Each agency has employed a different TC identification and tracking procedure to determine the number of TCs produced by their model within each ensemble member in each year of the hindcast.

In this section, the ability of each model to produce an environment consistent with observations is examined in the context of environmental parameters related to TC genesis. The TC tracking methods are presented and their performance when applied to the respective model hindcast ensembles is evaluated.

4.1. A comparison of dynamical seasonal tropical cyclone predictions for the Australian and Western Pacific regions

4.1.1. TC tracking in POAMA

Dynamical model POAMA [18] is comprised of 30-member ensemble and 31-year hindcast (1980–2010). Realisations initialised on 1 October each year (i.e. prior to start of the Southern Hemisphere TC season) are used to evaluate model's performance. Each realisation provides 9 months of daily global atmospheric environmental fields at approximately $2.5^\circ \times 2.5^\circ$ resolution.

TCs are identified and tracked using 'Okubo-Weiss-Zeta Parameter' (OWZP) scheme [19]. In brief, regions of low deformation vorticity (large OWZP) at 850- and 500-hPa levels which are vertically coherent and are sustained for an appreciable duration (at least 48 hours) are identified. Where such regions occur in presence of small vertical wind shear and large lower tropospheric humidity, local environment is considered conducive to imminent TC genesis or TC maintenance.

OWZP scheme is applied to 9-month daily POAMA data for each ensemble member and year individually. Statistics for TC-like disturbances for each member are then averaged together to give ensemble mean statistics for each year.

4.1.2. TC tracking in JMA/MRI-CGCM

The seasonal JMA/MRI-CGCM [20] is comprised of a 10-member ensemble, and the same 31-year hindcast period as POAMA is utilised in this analysis. The realisation for the forecast period beginning 1 November each year is used; each realisation provides daily global atmospheric environmental fields covering the November-April period at approximately $1.875^{\circ} \times 1.875^{\circ}$ resolution.

TCs are identified and tracked using a method similar to that outlined by [21], whereby grid points with 850 hPa relative vorticity less than a threshold value are identified and the sea-level pressure minimum within the surrounding grid points is denoted as the centre of the possible TC. A warm-core is required of these possible TCs: the thickness between 500 and 200 hPa must exceed that of the local environment, and wind speed at 850 hPa must be greater than that at 200 hPa. A full description of this method can be found in [22].

For a possible TC to be considered, it must last longer than 2 days and be equatorward of 30°S.

This scheme is applied to daily JMA/MRI-CGCM data for the November-April period for each ensemble member and year individually. Statistics for TC-like disturbances for each member are then averaged together to give ensemble mean statistics for each year.

The TC identification and tracking method is basin dependent. The method is applied to the global atmospheric fields using a variety of 850 hPa relative vorticity thresholds. For each basin, the observed climatological number of TCs is compared to the ensemble mean hindcast climatological value for each low-level vorticity threshold; the closest value is then selected for each basin.

4.2. Model environment

The ability of the models to represent the large-scale environment in which the TCs form has been demonstrated for 850 hPa relative vorticity vertical and troposphere-deep (850–200 hPa) vertical wind shear [23]. The results discussed here use a threshold value of $4.5 \times 10^{-5}\ s^{-1}$ for the Australian region and $7.5 \times 10^{-5}\ s^{-1}$ for the South Pacific region.

During November-December-January (NDJ), both models realistically capture the variability in low-level vorticity near the equator in the western Pacific with correlation values exceeding 0.8 is places. In the tropical Australian region the models do much less well.

The model drift associated with longer lead times is clearly evident in the February-March-April (FMA) season plots, with correlation values of low-level vorticity reduced from NDJ for both models.

Similar statements apply for both models in terms of the vertical wind shear.

4.3. Seasonal TC prediction

Ensemble mean variability in the number of TCs in the SH TC season (November-April; NDJFMA) is shown in **Figures 8** and **9**, along with the observed number of TCs from the BoM SH TC dataset [7].

http://dx.doi.org/10.5772/64029

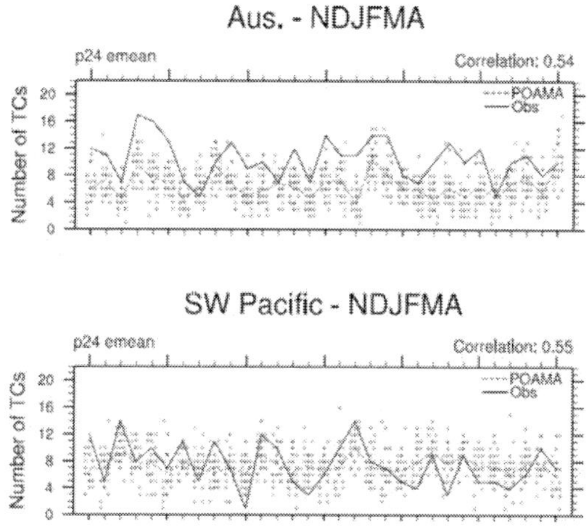

Figure 8. Time series of annual (NDJFMA) number of TCs in POAMA for the Australian (top panel) and South Pacific (bottom panel) regions. TC numbers are shown for observations (solid black line) and ensemble mean (dashed grey line). Ensemble members are shown as coloured circles.

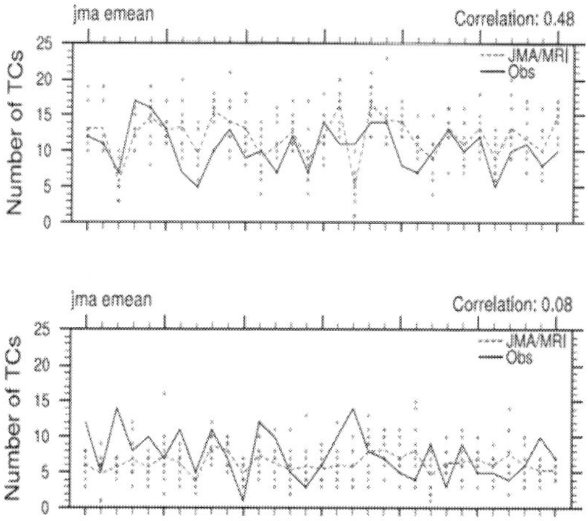

Figure 9. Time series of annual (NDJFMA) number of TCs in JMA/MRI-CGCM for the Australian (top panel) and South Pacific (bottom panel) regions. TC numbers are shown for observations (solid black line) and ensemble mean (dashed grey line). Ensemble members are shown as grey circles.

In the Australian region POAMA underestimates the number of TCs throughout the hindcast period, suggesting a deficiency in the model's ability to produce TC-like disturbances in this region. This is not the case in the western South Pacific. In both basins, some of the inter-annual variability is captured by POAMA, yielding a correlation values with observations of ~0.55.

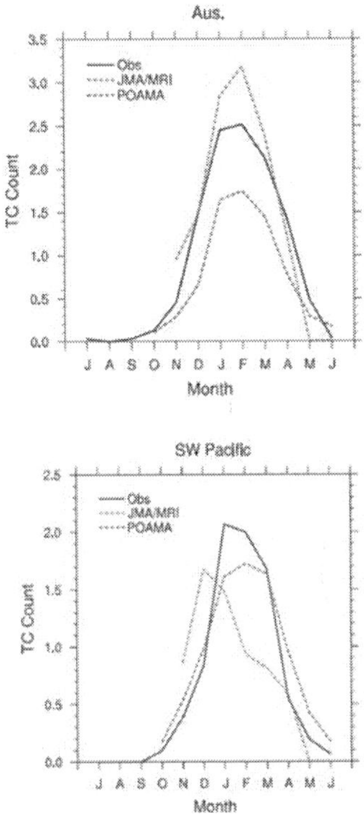

Figure 10. Climatological number of TCs as a function of month for observations (solid black line), POAMA (blue dashed line) and JMA/MRI-CGCM (red dashed line) in the Australian (top panel) and South Pacific Ocean (bottom panel) regions.

By design JMA/MRI-CGCM yields annual totals of TCs for NDJFMA close to climatology for both basins and in the Australian region a similar degree of variability to POAMA is captured, demonstrated by a correlation value of 0.48. In the South Pacific JMA/MRI-CGCM fairs less well at capturing the variability.

Ensemble mean monthly TC climatologies for each basin and model are shown compared with observations in **Figure 10**.

Both models capture the monthly variability in the Australian region well, although neither model represents the peak value correctly. In the South Pacific, POAMA performs well, however, JMA/MRI-CGCM peaks a month too early and drops off too quickly.

In summary, POAMA and JMA/MRI-CGCM both represent the large-scale environment relevant to TCs reasonably well, although possible deficiencies exist in the Australian region. The monthly TC climatologies in both models are reasonably realistic. Both models capture some of the inter-annual variability in the Australian region, although POAMA performs better in the South Pacific. Probabilistic NDJFMA TC number predictions both models, evaluated over the 31-year hindcast, show skill over random chance.

With further development of dynamical climate models and improving of their skill it is expected that both statistical and dynamical models will be used in operational TC seasonal prediction in the Australian and South Pacific regions, to complement each other.

5. Conclusions

Historically, multi-hazards associated with TCs (destructive winds, storm surges, torrential rain and related flash-flooding) had significant impacts on population and coastal infrastructure of Australia and island countries of the Indian and Pacific oceans. Improved forecasting of TC seasonal activity is an important part of the Climate Risk Early Warning System (CREWS) for improving resilience of the society to potentially destructive impacts of TCs. Currently, a statistical model-based prediction of TC activity in the coming season is used for operational seasonal forecasting in the Australian region and the South Pacific Ocean by the Australian Bureau of Meteorology, the National Institute for Water and Atmospheric Research (NIWA) in New Zealand and the Guy Carpenter Asia-Pacific Climate Impact Centre (GCACIC) at the City University of Hong Kong.

In this chapter, a possibility of improving the accuracy of TC seasonal forecasting using advanced statistical model-based approach (e.g. support vector regression methodology) was demonstrated. Moreover, it was shown that dynamical (physics-based) climate models have potential for skilful seasonal TC forecasting. Transition from a statistical to a dynamical prediction system will ultimately provide more valuable and applicable climate information about TC seasonal variability, which can inform decision making, responses and adaptation in Australia and Pacific and Indian Ocean island countries.

Acknowledgements

The research discussed in this paper was conducted with support of the Pacific Climate Change Science Program and Pacific-Australia Climate Change Science and Adaptation Planning program. D. Jones, R. Fawcett, J. Chan, R. de Wit, J. Apajee, Y. Wang, K. Shelton, A. Charles, T. Nakaegawa, J. Wijnands and G. Qian contributed to this research.

Appendix: Brief description of the developed statistical models

A. Linear discriminant analysis (LDA) model

Linear discriminant analysis (LDA) model is currently used as an operational model for TC seasonal prediction by a number of organisations including the Australian Bureau of Meteorology which utilise NIÑO3 and SOI indices as LDA models' inputs [7]. Examining prospects for improving skill of operational TC seasonal forecasting, Kuleshov et al. [15] demonstrated that 5VAR index performs better than NIÑO3 and SOI. Consequently, the LDA model for annual total occurrences of TCs in the Australian region (*AR*) has been modified to use 5VAR and also incorporate time trend variable (*T*) as predictors in the region:

$$AR = \beta_0 + \beta_1 T + \beta_2 5VAR + \epsilon$$

where $\epsilon N(0, 2)$. For a detailed mathematical description of the developed LDA model, see [15].

B. Support sector regression (SVR) model

Support vector regression (SVR) has been identified as a skilful machine learning algorithm for application to TC seasonal prediction [9]. Using non-parametric and non-linear regression approach, annual total number of TCs expected to be formed in the coming season (*Y*) has been generated using nine variables as the model's input. Selected input variables (X_1–X_9) were the following indices: X_1, Dipole mode index; X_2, NIÑO4; X_3, NIÑO3.4; X_4, NIÑO3; X_5, NIÑO1.2; X_6, El Niño Modoki index; X_7, 5VAR index; X_8, multivariate ENSO index; and X_9, SOI. For a detailed mathematical description of the SVR model, see [17].

Author details

Yuriy Kuleshov

Address all correspondence to: y.kuleshov@bom.gov.au

1 Bureau of Meteorology, Melbourne, VIC, Australia

2 School of Mathematics and Statistics, The University of Melbourne, Parkville, VIC, Australia

3 School of Science, Royal Melbourne Institute of Technology (RMIT) University, Melbourne, VIC, Australia

4 Faculty of Sciences, Engineering and Technology, Swinburne University of Technology, Melbourne, VIC, Australia

References

[1] Y. Kuleshov, S. McGree, D. Jones, A. Charles, A. Cottrill, B. Prakash, T. Atalifo, S. Nihmei and F. L. S. K. Seuseu, "Extreme weather and climate events and their impacts on island

countries in the Western Pacific: Cyclones, floods and droughts", Atmospheric and Climate Sciences, 4, 2014, 803–818. doi: 10.4236/acs.2014.45071.

[2] Y. Kuleshov, L. Qi, R. Fawcett and D. Jones, "On tropical cyclone activity in the southern hemisphere: Trends and the ENSO connection", Geophysical Research Letters, 35, 2008, L14S08, doi: 10.1029/2007GL032983.

[3] IPCC AR5, Intergovernmental Panel for Climate Change (IPCC) Fifth Assessment Report, 2013 (https://www.ipcc.ch/report/ar5/wg1/).

[4] Y. Kuleshov, "Climate change and Southern Hemisphere tropical cyclones", International Initiative, in U.C. Mohanty et al. (Ed.), "Monitoring and Prediction of Tropical Cyclones over Indian Ocean and Climate Change", Capital Publishing Co & Springer Publications, ISBN 978-94-007-7719-4, 2013, pp. 18–32, doi: 10.1007/978-94-007-7720-0_2.

[5] W. M. Gray, "Hurricanes: their formation, structure and likely role in the tropical circulation", In: D.B. Shaw (Ed.), Supplement to Meteorology Over the Tropical Oceans, 1979, James Glaisher House, Grenville Place, Bracknell, Berkshire, RG 12 1BX, pp. 155–218.

[6] N. Nicholls, "A possible method for predicting seasonal tropical cyclone activity in the Australian region." Monthly Weather Review, 107, 1979, 1221–1224.

[7] Y. Kuleshov, L. Qi, R. Fawcett and D. Jones, "Improving preparedness to natural hazards: Tropical cyclone prediction for the Southern Hemisphere", In J. Gan (Ed.), Advances in Geosciences, Vol. 12, "Ocean Science", World Scientific Publishing, Singapore, 2009, pp. 127–143.

[8] K. S. Liu and J. C. L. Chan, "Interannual variation of Southern Hemisphere tropical cyclone activity and seasonal forecast of tropical cyclone number in the Australian region", International Journal of Climatology, 2010, 33(2), 190–202, doi: 10.1002/joc.2259.

[9] J. S. Wijnands, K. Shelton and Y. Kuleshov, "Improving the operational methodology of tropical cyclone prediction in the Australian and the South Pacific Regions", Advances in Meteorology, 2014, 2014, Article ID 838746, http://dx.doi.org/10.1155/2014/838746.

[10] G. J. Holland, "On the climatology and structure of tropical cyclones in the Australian / southwest Pacific region: I. Data and tropical storms", Australian Meteorological Magazine, 32, 1984, 1–15.

[11] M. Broomhall, I. Grant, L. Majewski, M. Willmott, D. Jones and Y. Kuleshov, "Improving the Australian tropical cyclone database: Extension of GMS satellite image archive", In Y. Charabi (Ed.), Indian Ocean Tropical Cyclones and Climate Change, Springer, NY, 2010, pp. 199–206, doi: 10.1007/978-90-481-3109-9_24.

[12] A. Dowdy and Y. Kuleshov, "An analysis of tropical cyclone occurrence in the Southern Hemisphere derived from a new satellite-era dataset", International Journal of Remote Sensing, 33(23), 2012, 7382–7397, doi: 10.1080/01431161.2012.685986.

[13] Y. Kuleshov, R. Fawcett, L. Qi, B. Trewin, D. Jones, J. McBride and H. Ramsay, "Trends in tropical cyclones in the South Indian Ocean and the South Pacific Ocean", Journal of Geophysical Research, 115, 2010, D01101, doi: 10.1029/2009JD012372.

[14] H. A. Ramsay, L. M. Leslie, P. J. Lamb, M. B. Rickman and M. Leplastrier, "Interannual variability of tropical cyclones in the Australian region: Role of large-scale environment", Journal of Climate, 21, 2008, 1083–1103.

[15] Y. Kuleshov, Y. Wang, J. Apajee, R. Fawcett and D. Jones, "Prospects for improving the operational seasonal prediction of tropical cyclone activity in the Southern Hemisphere", Atmospheric and Climate Sciences, 2(3), 2012, 298–306, doi: 10.4236/acs. 2012.23027

[16] A. Dowdy, "Long-term changes in Australian tropical cyclone numbers", Atmospheric Science Letters, 15(4), 2014, 292–298, doi: 10.1002/asl2.502.

[17] J. S. Wijnands, G. Qian, K. Shelton, R. J. B. Fawcett, J. C. L. Chan and Y. Kuleshov, "Seasonal forecasting of TC activity in the Australian and the South Pacific Ocean regions", Mathematics of Climate and Weather Forecasting, 1, 2015, 21–42, doi 10.1515/ mcwf-2015-0002.

[18] S. Langford, H. H. Hendon and E.-P. Lim, "Assessment of POAMA's predictions of some climate indices for use as predictors of Australian rainfall", CAWCR Technical Report, 031, 2011, 60 pp.

[19] K. J. Tory, R. A. Dare, N. E. Davidson, J. L. McBride and S. S. Chand, "The importance of low-deformation vorticity in TC formation", Atmospheric Chemistry and Physiscs, 12, 2012, 17539–17581.

[20] T. Yasuda, Y. Takaya, C. Kobayashi, M. Kamachi, H. Kamahori and T. Ose, "Asian monsoon predictability in JMA/MRI seasonal forecast system", CLIVAR Exchanges, 43, 2007, 18–20.

[21] F. Vitart and T. N. Stockdale, "Seasonal forecasting of tropical storms using coupled GCM integrations", Monthly Weather Review, 129, 2001, 2521–2537.

[22] Y. Takaya, T. Yasuda, T. Ose and T. Nakaegawa, "Seasonal prediction of mean location of Typhoon formation", Journal of the Meteorological Society of Japan, 88(5), 2009, 799–812.

[23] A. Charles, K. Shelton, T. Nakaegawa, H. Hendon and Y. Kuleshov, "Prediction of tropical cyclone activity with coarse resolution global climate models", Proceedings of the 20th International Congress on Modelling and Simulation (MODSIM2013), Adelaide, Australia, 1–6 December, 2013, pp. 2555–2561.